T0255388

The Cambridge Nature Study Series

General Editor: HUGH RICHARDSON, M.A.

POND PROBLEMS

Pond in the grounds of Leighton Park School.

POND PROBLEMS

BY

ERNEST E. UNWIN, M.Sc.

HOUSE MASTER AND SCIENCE MASTER AT LEIGHTON PARK SCHOOL, READING
FORMERLY SCIENCE MASTER AT ACKWORTH AND BOOTHAM SCHOOLS
AND DEMONSTRATOR IN BIOLOGY IN THE UNIVERSITY OF LEEDS

Cambridge :
at the University Press
1914

CAMBRIDGE
UNIVERSITY PRESS

University Printing House, Cambridge CB2 8BS, United Kingdom

Published in the United States of America by Cambridge University Press, New York

Cambridge University Press is part of the University of Cambridge.

It furthers the University's mission by disseminating knowledge in the pursuit of education, learning and research at the highest international levels of excellence.

www.cambridge.org
Information on this title: www.cambridge.org/9781107642249

© Cambridge University Press 1914

First published 1914
First paperback edition 2014

A catalogue record for this publication is available from the British Library

ISBN 978-1-107-64224-9 Paperback

PREFACE

IT has been my privilege to work for some three years with Mr Unwin as my colleague at Bootham School. Inhabiting a study next to his lecture room I have heard fragments of many of these lessons and I know for myself how keen has been the interest evoked, how eager the little crowd round the table afterwards. It has been a revelation to me to see how standards of knowledge appropriate to a university professor have captured the confidence of the youngest class of schoolboys. Now that even the kindergarten dabbles in pond life it is necessary to appeal for far higher standards of thoroughness if the same topics are still to rivet attention at an older age.

This book is really about environment, natural selection and evolution, even if such long words are avoided in the text. The lessons have been tested in detail, and information about apparatus and materials will be useful both to teacher and pupil in arranging for practical work.

Mr Unwin has had great experience as a teacher, having been for four years Demonstrator in Zoology at Leeds University, Science Master at Ackworth, Bootham and Leighton Park Schools, master of a preparatory school, as well as lecturer on Nature Study to adult audiences of teachers and others. He is also his own artist and the illustrations in this book, almost without exception, are from his pen or camera.

HUGH RICHARDSON.

BOOTHAM SCHOOL, YORK,
May 1914.

CONTENTS

LIST OF ILLUSTRATIONS

The illustrations (except Figs. 7 and 47) are reproduced from the author's own drawings and photographs.

INTRODUCTION

(ADDRESSED TO THE TEACHER)

THE study of environment and its effect upon living
things is one of the most interesting branches of
the science of life. It is the study of the living organism
in connection with its living and non-living surroundings.
Methods of study which apply to beetle larvæ and to
caddises can afterwards be applied to the study of man
in the slum or garden-city. Insects, so numerous and
so changeable, offer themselves as good subjects for
such a study, and among the insects the aquatic forms
are the most striking and interesting. The ease with
which aquatic insects can be kept in captivity is an
additional reason for their important place in any
scheme of Nature Study.

Nature Study has sometimes been too vague, and
it sometimes degenerates into talks by the teacher and
the pupils doing very little, if any, definite practical
work. For very young children this matters less, so
long as curiosity and wonder are not destroyed; but
for older boys and girls the subject needs a more
scientific basis. The spirit of Darwin should be brought
into our work in its fourfold character of curiosity, of
observation, of experiment, and of the power to draw
conclusions from the facts obtained by the observations
and experiments.

Nature Study should be regarded as research, and the work so planned that the pupils have to use their own powers of observation and reflection. One of the best ways to do this is by means of questions which lead from one point to another. The questions can only be answered by careful observation. Thus unconsciously the scientific method of approaching any problem will be learnt and Nature Study will become important as a training ground for the study of the problems of Physics, Chemistry, Biology and Sociology.

This method of having definite questions to answer (whether written on the blackboard or printed in a book) releases the teacher and enables him to go round the class helping, suggesting, and encouraging the pupils in their research.

No method is perfect and the questions should not be used in too rigid a manner. A good teacher will take the lesson into his own hands, some questions will be discussed and answered verbally, new questions will continually arise, details not easily obtainable by the class will be filled in, and at the end of any piece of work the results will be summarised by the class and by him.

This question method was learnt and practised by the author under the guidance of Professor L. C. Miall, F.R.S., (then) Professor of Biology in the University of Leeds, and put into further practice whilst teaching at Ackworth, Bootham and Leighton Park Schools. The author wishes to acknowledge the constant inspiration and help given by Professor Miall during the years spent in Leeds, first under him as a student and then as a colleague. This book is the outcome of attempts to put Professor Miall's work on Aquatic Insects into a form suitable for use in the classroom; indeed it is expected that his book *The Natural History of Aquatic Insects*[1] will be at hand and used as directed herein.

In examining and describing any new animal or structure, it will be found helpful to have some scheme to guide the observation in a right sequence. Such a scheme is given below.

    I. Kind of thing it is.

    II. Position (this is for description of organs).

    III. Form, Size, Colour.

    IV. General Structure.

    V. Minute Structure.

    VI. Function.

The teacher is also directed to the naturalist's four chief questions which Professor J. Arthur Thomson has kindly allowed me to reprint from his pamphlet of suggestions to teachers[2].

The first question is—" *What is this?* ": an inquiry into *form* and *structure*. What is this living creature in itself and in its parts? What is it as we see it with our own lenses only, and as we see it when we put other lenses in front of ours? What is it as a thing by itself, and when compared with its fellows and kindred?

The second question is—" *How does this act?* ": an inquiry into *habits* and *functions*. How does this living creature behave as it does? What is its business? How does it keep agoing and set

---

[1] Published by Macmillan at 3s. 6d.

[2] *Some suggestions to teachers for seasonal Nature Study.* This is now out of print but the suggestions have been incorporated in a Board of Education Memorandum on *Nature Study and the Teaching of Science in Scottish Schools* [Cd 4024], Wyman & Sons, 3d. Teachers are also directed to a fuller treatment of the subject in *Darwinism and Human Life*, J. Arthur Thomson (Melrose, 5s.), page 6.

other creatures like itself agoing ? How does it get on ? What is the "particular go" of it ?

The third question is—"*Whence is this?*": an inquiry into *development* and *history*. Where did this living creature come from ? How did it begin ? What was it like when it was young ? What are the chapters in its growth and life-history ? What is known of the history of its race ?

The fourth question is—"*How has this come to be as it is?*": an inquiry into *causes*. What factors have led to this living creature being what it is, where it is, as it is ; in short, what have been the factors in its evolution ?

It need hardly be.said that these are not questions for children. They are the fundamental questions of the science of biology, which is not for children. But they are stated here because they help greatly to keep our own minds—as teachers—in good order.

This book is intended for the lower forms of Secondary Schools, for Preparatory Schools and for the higher standards of Elementary Schools.

CHAPTER I

HOW TO OBTAIN SPECIMENS

THIS book is planned upon the assumption that the insects mentioned in it are available for observational work by the class; it is therefore necessary to give

Fig. 1. The capture of specimens. A pond near Ackworth.

directions which will enable the specimens to be obtained. There are two methods which can be adopted:

(1) the capture of the insects by the teacher and class, and (2) buying from recognised dealers. The first method is much the better; and, in most cases, the choice of "Pond Problems" as a Nature Study Course will be made by a teacher having access to convenient ponds and streams in the neighbourhood. At the same time it should be remembered that most of the insects can be got from dealers (see Appendix), and it is possible to keep a good supply of insects in aquaria.

Fig. 2. The capture of specimens. A pond near Ackworth School.

The first thing to think about is a suitable net to catch the insects. Everyone has his own ideas, and the net I have found most useful is shown in Fig. 3. It is made of brass, with brass gauze (10 squares to 1 linear inch), to fit on to the end of a walking-stick and when not in use the net can be put in a pocket. Other nets will do equally well, an iron ring and canvas is quite sufficient, or suitable nets can be bought from dealers in Nature Study material (see Appendix for details).

For carrying home the finds small tin boxes (tobacco tins) are the best. Glass bottles with wide necks holding 6 or 8 ozs. and glass specimen tubes (3″ by 1″) are useful. A pair of broad-tipped forceps will be helpful in removing specimens from the net. Most insects travel much better in damp air than in small supplies of water which soon get warm in the pocket and exhausted of oxygen.

Fig. 3. Brass collecting-net made to fit on to a walking-stick and when not in use to go into a side pocket.

Be careful not to mix the specimens or on arrival at home you will have evidence of the struggle for existence.

The places where aquatic insects can be found are many and varied. Rainwater tanks and butts, drinking troughs especially in fields, slow muddy streams, rapid stony streams and all kinds of ponds.

For help in obtaining specimens it will be convenient to describe, in alphabetical order, the insects mentioned in this book, some rough means of identification and likely places for their capture.

Insect.	Description.	Where?	When?
Alderflies (*Sialis*)	Blackish fly, 4 clear wings	Near slow streams and ponds	Summer
,, larvæ	[1]Miall, Fig. 86	In mud of above	All the year
Beetles			
Dytiscus	Miall, Figs. 8 and 9. Unwin[2], Fig. 12	In ponds	,,
,, larvæ	,, Fig. 4. Unwin, Figs. 17 and 18		,,
Great Water Beetle (*Hydrophilus*)	,, Fig. 18	In weedy ponds	,,
Hydrobius	Black. Miall, Fig. 19	In very weedy ponds crawling about the duckweed	,,
Helophorus	Brown		
Whirligig	Miall, Fig. 1	Surface of ponds	,,
Bloodworms, larvæ of *Chironomus*	See under Gnats		
Caddises			
larvæ	Various. See Miall, Fig. 82. Unwin, Fig. 31	Ponds and streams	All the year
flies	Brownish moth-like	Near ponds and streams	Summer
Chironomus }	See under Gnats below		
Culex }			
Dragonflies larvæ	Several kinds, Miall, Fig. 95	Near streams and ponds	Summer
	1. Small forms, with three plates at end of body. *Agrionidæ* family. Unwin, Fig. 24		
	2. Larger, broad flat abdomen, shorter than the hind legs. *Libellulidæ*. Unwin, Fig. 24	Ponds	All the year
	3. Larger, with larger abdomen, longer than hind legs. *Æschnidæ*. Miall, Fig. 96. Unwin, Fig. 24		
Diza	See under Gnats		
Eristalis	See Rat-tailed Maggot		
Gnats			
Chironomus (Harlequin-fly)	Miall, Fig. 43. Fly rests with fore legs raised	Swarms near water, on windows and fences	Summer and autumn
larvæ (bloodworm)	Miall, Fig. 37	Water-troughs, muddy streams, ponds	All the year
pupæ	Miall, Fig. 39. Unwin, Fig. 42 }		

Culex (Biting gnat)	Miall, Fig. 27. Fly rests with hind legs raised	Windows and walls, or near ponds	Summer and autumn chiefly
larvæ	Miall, Fig. 22. Unwin, Fig. 19	Butts, tanks, ponds and streams	Abundant in summer
pupæ	„ „ 24. „ „ Fig. 20		Summer
egg-rafts	„ „ 30. „ „ Fig. 46		
Dixa larva	Small blackish larva. Miall, Fig. 47	Ponds	,,
Tanypus larva	Bloodworm-like, more vigorous, darker colour. Miall, Fig. 45	Ponds	,,
Mayflies	Moth-like flies with 3 long-tail filaments	Near ponds and streams	Early summer
Larvæ			
1. Swimming (*Chloeon*)	Feathered tail filaments, cylindrical body, small, lightly built with legs small and thin. Unwin, Fig. 26	Ponds	All year—abundant in summer
2. Burrowing (*Ephemera vulgata*)	Light brownish body cylindrical, large, broad legs for burrowing. Unwin, Fig. 30. Miall, Fig. 93	Streams, especially at sandy bends	All the year
3. Rock-clinger (*Baetis*)	Flattened body. Miall, Fig. 94	On the under-side of stones in a rapid stream	,,
Phantom larva (*Corethra*)	Small transparent larva. Miall, Fig. 31	Shady ponds and ditches	Summer
Pond-skaters	Miall, Figs. 102, 103, 104, 105, 106	Surface of ponds and streams	,,
Ptychoptera larva	Whitish maggot with long thread-like tail, Miall, Fig. 56	Shallow muddy ponds or sandy streams	All the year
Rat-tailed Maggot (*Eristalis*)	Similar to above but larger body. Miall, Fig. 70	Edges of ponds with much decaying matter	,,
Sialis, see Alderfly			
Simulium larvæ and pupæ	Miall, Figs. 59 and 65	Rapid streams	All the year
Stone-flies and larvæ	Miall, Figs. 87 and 89	Stony streams along with *Baetis*	Summer. All the year. Flies in early summer
Stratiomys larvæ	Miall, Fig. 67	Shallow weedy ponds	All the year
Water-boatmen, *Corixa* and *Notonecta*	Miall, Figs. 108 and 110. Unwin, Figs. 29, 14 and 15	Ponds	,,
Water-measurers	See Pond-skaters		
Water-scorpions	Miall, Fig. 107. Unwin, Fig. 22	Weedy ponds	,,

[1] Miall, when used in this way, refers to *Nat. Hist. of Aq. Insects* (Miall).
[2] Unwin, when used in this way, refers to this book—*Pond Problems.*

It must be remembered that ponds, streams, and rainwater butts vary enormously and the question of *Eyes and No-Eyes* or *Man sieht nur was man weiss* will have some effect upon the extent of the capture.

In general aquatic larvæ can be found all the year, but of course more plentifully in summer. Pupæ and winged forms in the summer. During severe cold many aquatic insects hibernate in the mud.

For other insect material and for apparatus and methods, see Appendix.

Aquaria.

The next thing after the capture and return home is the right housing of the creatures. Many books have been written about the right method of keeping aquatic animals in captivity[1], so that only a few general hints will be given here.

The simpler the aquarium the better.

Never use goldfish globes. The proportion of air surface is too small, the curved sides are bad for observations, and too much light is allowed to get to your artificial pond. Bell-jars are also open to the last two disadvantages.

Use shallow earthenware dishes. Pie-dishes and large saucers are excellent or the shallow earthenware dishes glazed white inside, sold in nests of 3 or 4.

Keep different insects in different dishes or, if you mix them, see that kinds harmless to each other live together. Experience will teach this.

---

[1] *School Aquarium*, Miss C. V. Wyss. Sch. N. St. Union. Price 2½d., 1s. a dozen. *The Book of Nature Study* (Caxton Pub. Co.), Vol. II, *Life in Ponds and Streams*, Furneaux. *Fresh-water Aquaria*, Rev. G. C. Bateman (Upcott Gill, 3s. 6d.).

A little weed should be put in each dish.

Cover the dishes with glass plates to keep down evaporation and keep out the dust. This is not always possible nor always necessary.

Keep one large tank for a collection of animals. Such a tank can be made of wood and glass, or a large glass tank can be adapted. Such tanks can

Fig. 4. Naturalist's work-table, showing compound and dissecting microscope and various kinds of aquaria from a saucer to a large tank.

be got from electric supply firms, and are made for accumulator tanks.

For such an aquarium any or all of the following will do.

Plants. Water starwort (*Callitriche*), American pondweed (*Elodea* or *Anacharis*), a little duckweed and, rooted in clean sand or held down by rocks or pieces of lead piping, *Potamogeton* and water crow-foot.

Animals. A few snails (*Limnæa* and *Planorbis*), caddises, gnat larvæ, water-fleas, one or two dragonfly larvæ, mayfly larvae (*Chlœon*), a water-spider (this may have to be removed).

Fig. 5. Glass tank in use   Watching the breathing of a water-beetle. See Chapter IV, p. 21.

Clean river sand and a few rocks are also needed.

Small rectangular glass tanks, as in Fig. 5, will be needed for observational work, but they are not suitable for most insects as permanent abodes. (See Appendix.)

CHAPTER II

WHAT IS AN INSECT?

I. An insect's near relations.

Material for each pupil[1].

Boiled shrimp, spider, a large insect (bee, beetle or wasp), centipede, pocket lens.

1. Write down carefully all the things that a shrimp, insect, spider and centipede have in common.

2. Write down the more striking differences between them. Try to find important differences.

All these creatures belong to one of the largest groups of animals. They are all Arthropods, that is jointed legged animals, and form one group of the non-backboned animals. You will have noticed that their bodies are jointed as well as their legs. Arthropods are all covered with a special kind of "skin" which is sweated out by the true skin, and which hardens to form an outer armour or skeleton. This horn-like skin is called **chitin**.

The differences between the four animals will form a basis for a rough classification of the Arthropods. Shrimps with many legs, 2 pairs of feelers, belong to the **Crustacea** which mostly live in water and breathe by gills. Spiders with head and thorax united, 8 legs and no antennæ (feelers) belong to the **Arachnida**. Centipedes with their many divisions and many legs and 1 pair of antennæ belong to the **Myriopoda**, and wasps with three well-marked divisions of the body

---

[1] See Appendix for Material and Apparatus.

(head, thorax and abdomen), 6 legs, 4 wings (some insects have 2 and some none) and 1 pair of antennæ belong to the **Insecta**.

3. Make drawings of the shrimp, spider, centipede and insect to show these points. Carefully name your drawings.

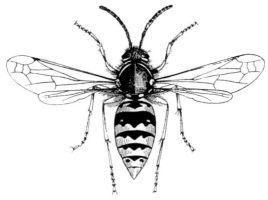

Fig. 6. Dorsal view of a wasp as an example of a perfect insect. × $\frac{5}{2}$.

II. The structure of an insect.

A. **Material for each pupil.**

Living cockroaches, Petri dishes, pocket lens, a small quantity of flour.

The cockroaches can be conveniently studied alive by imprisoning them between the upper and under glass of a Petri dish. In this way both upper and under surface can be seen.

4. Watch the movements of the antennæ and mouth parts.

5. Sprinkle a little flour over the head and antennæ. Watch the cleaning process. Describe carefully how this is done.

6. Write down the reasons for saying a cockroach is an insect.

7. Draw a top view of the cockroach natural size.

B. **Materials.**

Caterpillars (preserved) for each pupil. Microscopes and mounting material. See Appendix.

8. Examine a dead caterpillar. Make out the head, thorax with jointed legs and abdomen with stump-like pro-legs.

9. Find along each side of the body a row of small oval brown marks. Note carefully on which segments they are found.

Fig. 7. Side view of a caterpillar (death's-head moth) showing spiracles. Nat. size.

10. Make a careful side-view drawing of the caterpillar twice natural size, and put in all the things already discovered.

These oval marks are called **Spiracles**. Insects do not breathe through their mouths but through these spiracles, which are breathing holes.

11*. Cut out a piece of the skin from the side of a caterpillar so as to include at least one spiracle. Scrape away muscle and fat

from the inside and mount on a slide in dilute glycerine. Examine
upon 1″ power of microscope.

12*.   Make a drawing of the piece of skin and spiracle.

13*.   Into what do these spiracles open ?

*Preparations for answering these questions can be
prepared beforehand by the teacher or bought from
dealers in microscopic slides[1].

Take a few dead smooth-skinned caterpillars, cut
them into right and left halves longitudinally. Remove
the food canal and fat carefully. Place the halves in a
bottle of 5 % caustic potash for a day or for 10 mins.
or so if the caustic potash is made hot. Wash in water
and mount in dilute glycerine. Or better still, the pieces
of skin can be made into permanent mounts by using
Canada balsam. (For procedure see Appendix.)

It has been found that 4 or 5 microscopes are sufficient for a class of 20, when this question method is in
use. The pupils do not work at the same rate and it
is possible, by having 5 slides of the same subject, to
get the microscopic work done by everybody without
confusion. If a lantern microscope is available and
readily adjusted, a large class can see, describe and
draw the object at the same time.

As a further help in giving a complete answer to
Question 13, the teacher could dissect a large cater-
pillar. (For method see Miall's *Injurious and Useful
Insects*, p. 51[2].)

It will now be evident that insects do not possess
lungs but that the spiracles open into fine air-tubes which
branch all over the body. In a large number of animals,
in ourselves for instance, the blood is used as the carrier

---

[1] For names of dealers see Appendix.

[2] *Injurious and Useful Insects*, Miall (Bell, 3s. 6d.).

and distributor of the oxygen of the air. The air is taken into a special changing bag where it comes into close contact with the blood, and then this blood is driven all round the body by the heart. But in insects the air itself goes to all parts of the body.

If a living wasp or bee can be caught and imprisoned in a test-tube with the end plugged with cotton-wool, the breathing actions can be shown. The renewal of air in the air-tubes and the driving of it into the fine branching tubes is brought about by the concertina-like contractions and expansions of the body.

14. If a living wasp or bee is available, count the number of breathing movements per minute.

15*. Examine a piece of air-tube under a high power. Notice the spiral thickening in the wall. Draw a short piece.

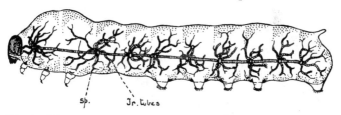

sp.   Jr. tubes

Fig. 8. Side view of the skin of a caterpillar, prepared so as to show the spiracles (sp.) and the tracheal tubes (tr. tubes) of one side. × 2.

This arrangement resembles the spiral wire some-times found in india-rubber gas tubing, and the wind-pipe of man has something similar to prevent collapse. It is important that the air-tubes of insects should be able to spring open when the pressure is relaxed so that fresh air can enter through the spiracles. The similar arrangement in man and insects has led to the use of the term **tracheal tubes** from the word **trachea** (windpipe).

C. **Materials.**

Slides prepared for Question 13, 2 crab's claws, 2 pieces of bone, 2 pieces of horn, 2 beetles or wing-cases of beetle. 2 glass jars, 10 % solutions of hydrochloric acid and caustic potash.

We are now going to discuss the question—" What is chitin ?"

Fig. 9. Spiny lobster (*Palinurus*) about to cast its skin.
(Photographed in one of the tanks at Port Erin Biological Station.)

16. Examine a piece of skin of an insect under the microscope. Slides of Question 13 will do—or a piece of the abdominal wall of a cockroach will answer admirably. Describe the character of the chitin, as you move the slide to see the whole of the piece of skin.

17. Make a drawing to show alternate thin and thick chitin. Explain the reason for this.

18. Prepare the following experiment. Take 2 glass sweet-jars, in one put 10 °/₀ caustic potash and in the other 10 °/₀ hydrochloric acid. In each jar place a crab's claw, pieces of bone and horn, and a beetle. Leave for 1 week and then examine the contents.

This experiment should show the wonderful resisting properties of chitin. The result of a similar experiment is given below.

Object	Condition at end of week	
	Caustic Potash	Hydrochloric Acid
Chitin (1) Claw (2) Insect	Unchanged Unchanged (except for internal soft parts which had disappeared)	Claw flexible. Lime salts removed No change
Bone	Very soft. Only fat left	Flexible. Lime salts removed
Horn	Disappeared	No change

Chitin is the only one not affected by either fluid. In **Crustacea** (crabs etc.) the chitin is encrusted with lime salts and this hardening substance is removed by the acid, but the chitin itself is unchanged. Although horn-like, it is not horn but some complex substance formed by the true skin. It has wonderful resisting properties, which are of great use to us in preparing the external features of insects for microscopic examination.

The presence of chitin outside the true skin and the fact that it is a secretion and therefore cannot grow, make a profound change in the ordinary procedure of growth in the **Arthropods**. A crab, a spider, a millipede and an insect, in order to grow have to cast off this armour of chitin when it is no longer large enough to hold the growing body, and a new and larger covering is made. This action is called moulting, and in consequence the growth appears to take place by abrupt changes rather than by steady imperceptible means. Before the old chitin is cast, a new layer of chitin is secreted by the true skin. This new layer is thin and

Fig. 10. The cast skin of the spiny lobster shown in Fig. 9. Notice the transverse crack in the middle of the back through which the lobster withdrew all its body.

flexible, and is often thrown into folds so as to stretch to a larger size after the moult.

Fig. 11 shows a stick insect (*Bacillus rossi*) hanging up to dry after a change of "skin."

The life of insects is governed by these periodic moults and great and vital changes, such as a fresh pattern of jaws, acquisition of wings, change of environment are effected at certain of these moults.

What is an insect? It is an animal, an invertebrate,

an arthropod with its segmented body grouped into three regions—head, thorax, abdomen. It has 3 pairs of jointed legs and is covered with chitin.

Insects are terrestrial and aërial, some are aquatic;

Fig. 11. The change of skin. A stick insect (*Bacillus rossi*) hanging up to dry after a moult.

they usually have wings when adult; they breathe air by means of tracheal tubes; their life is often divided into distinct periods.

CHAPTER III

THE PROBLEMS STATED

We have now studied the main points of insect structure, and we are in a position to appreciate the problems which face those insects which live for a longer or shorter period in water.

Why are there any problems? Insects are really land animals. They are terrestrial and aërial, and not aquatic. They are really air-breathers and do not belong to the water.

How do we know this? That this statement is true is very important, for the fact that insects do not belong to the water is the text for this book, for it is from that fact that all the problems come. What proofs have we that insects have invaded the water from the land, just as the whales and porpoises have? We know that living in water is a simpler business than living on land or in the air; we know that as we go down the animal kingdom we find more and more aquatic forms, and a very brief survey will convince us that the terrestrial habit and, still more, the aërial habit have been adopted by comparatively few animals. Why cannot we say that the aquatic insects are therefore more primitive, and the terrestrial insects have evolved from them?

The answer or the proof is a threefold one.

(1) **Geology** does not help us very much because the record is very incomplete, but what little is known clearly points to the ancestral forms being land creatures and air-breathers.

(2) **Classification** tells the same story. Anyone with but a casual acquaintance with water insects knows what a mixed set they are. A few beetles, a few bugs, a few early stages of two-winged flies, one or two caterpillars, and so the list proceeds. It is also quite clear that the simplest insects we know are not aquatic, and that the aquatic members of any order of insects are not more primitive than the terrestrial members. Again there is no connection between the aquatic members of one insect order and those of another. It would thus seem clear that not only have insects invaded the water, but that there have been several independent invasions.

(3) **Vestiges.** Just as in other branches of Biology, we can read the past from the vestiges left in the individual; just as in the horse, the splint bones tell of more toes than one in the ancestral horse, so in aquatic insects the presence of unused air-tubes, of plugged up spiracles speak unmistakably of a former terrestrial life.

19. Examine and draw the tracheal tube vestiges in a bloodworm—the larva of *Chironomus*. The bloodworm can be mounted so that the first two or three segments behind the head are seen under the low power.

Then besides this threefold proof, there is the further fact that *every* aquatic insect when in its perfect stage (imago) breathes by means of spiracles and air-tubes.

From these evidences we see clearly that insects have invaded the water from the land and air. They have done so to escape the severer competition on land. The struggle for existence has driven insects to seek shelter and food in the water just as the same struggle drives people to emigrate to Canada and Australia or to follow hazardous occupations in aeroplanes and sub-marines.

The problems are many.   The chief ones are :

(1)   The problem of breathing, solved in different ways by different insects.

Fig. 12.   *Dytiscus marginalis*.   Male and female beetles.
(The females are those with wings displayed.)

(2)   The problem of equilibrium and locomotion.

(3)   The escape from the water to the air.   Most aquatic insects only spend a portion of their life in the water, returning to land and air for a longer or shorter time before death.

These problems will be dealt with in the subsequent chapters.

CHAPTER IV

BREATHING. WATER-BEETLES AND WATER-BOATMEN

Material for each pupil or pair of pupils.

Living Dytiscus[1] *and dead specimen, living Hydrophilus or Hydrobius beetle, living and dead water-boatmen—both kinds (Corixa and Notonecta). Glass*

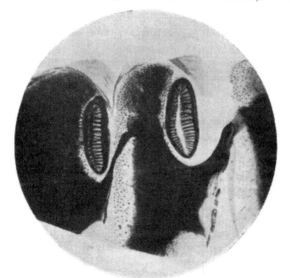

Fig. 13.   Photomicrograph of two abdominal spiracles
of the *Dytiscus* beetle.

tank, saucer, pocket lens, microscope.   Small pieces of plush.

Answer the following questions by observation on a living *Dytiscus* in a glass tank.   See Fig. 5.

[1] Any of the *Dytiscus* family would do: such as *Colymbetes* or *Agabus*.

20. How would you know it was an insect ?

21. How would you know it was a beetle ?

22. Is the beetle heavier or lighter than water ?

23. Can it remain below without holding on ?

24. Does it swim to the top or rise without effort ?

25. What part of the beetle reaches the surface first ?

26. Why does the beetle rise to the surface ?

27. Draw a quick sketch of its attitude whilst rising.

28. Describe carefully its position at the surface. Is there any special benefit in this ?

29. Draw this position from above and from the side.

30. Is there any visible breathing action ?

31. If you have time and opportunity, try to find out how long the beetle can remain below without coming up to breathe.

32. Has it any difficulty in diving below when disturbed at its breathing ?

33. Does it take any air below with it ?

34. Examine a dead specimen. Pin it out on a piece of sheet cork or cork linoleum with one wing and wing-case displayed or removed. If specimens are scarce, one or two prepared beforehand can be passed round. Describe the air cavity under the wing cases. How many spiracles can you see ? Make a drawing to show these things.

35. Examine the last spiracles; how do they differ from the others ? Cut one out, mount dry, and examine under the low power. Describe and draw it. Such a slide can be prepared beforehand.

36. What prevents water from entering these spiracles when in use at the surface ?

37. Drop water on a piece of plush. How does this illustrate the same point ?

If specimens of *Hydrophilus* or *Hydrobius* or *Helophorus* are available, compare them with *Dytiscus* in the following way.

38. Repeat Questions 20—33.

39. Does the antenna take any part in breathing ?

40. Why are these beetles called silver beetles ?

41. Take the beetle out of the water. What becomes of the silvery appearance ?

42. Examine the under surface of the beetle with a lens. Can you now explain why this beetle is silvery and *Dytiscus* not ?

43. Dip a small piece of plush beneath the surface of the water. How does this help to explain the above questions ?

For further details concerning the breathing of *Dytiscus, Hydrophilus* and *Hydrobius,* see *Natural History of Aquatic Insects,* pp. 75—81.

Examine living water-boatmen in tanks—dead specimens should be distributed.

44. Why are these creatures called insects ?

45. Examine and draw the head showing the piercing and sucking proboscis, which is characteristic of the Bug family (*Hemiptera*).

46. Compare the two kinds (*Notonecta* and *Corixa* with speckled back), noting shape, size, position of swimming etc.

47. Are they heavier or lighter than water ?

48. Is there any difference in this respect between the two kinds ? Can each stay below without holding on ?

49. Describe carefully the position of each when breathing at the surface, and if possible make drawings.

50. Are the oars (fringed legs) used at all in breathing ? Watch carefully before answering.

51. Examine *Notonecta*, either living or dead, out of the water. If living it must be held gently but firmly between finger and thumb. The abdomen is keeled. Describe what you see, and examine the grooves and fringes of hairs.

52. Watch *Notonecta* in the water : can you find out the use of these fringed hollows ?

53. Where would you expect the chief spiracles to be in *Notonecta* and why ?

54. Do either or both carry a supply of air below and, if so, where ?

55. How long can they stay below without renewing the supply of air ?

This problem of breathing has been faced and solved by the beetles and water-boatmen in a straightforward way. It has been similarly solved by other animals at different times in the history of the world. The whales, porpoises, and dolphins, like the aquatic insects, have invaded the water from the land. Thus insects are not alone in the "compulsory changing" made necessary by change of environment. The aquatic mammals (mentioned above) are air-breathers. They do not change their method of breathing, but merely modify it to suit the new circumstances as we do when we bathe. In the

Fig. 14.   Water-boatmen (*Notonecta*) breathing at the surface.

same way the water-beetles and water-boatmen still continue to breathe air, using spiracles and air-tubes, rising to the surface for fresh supplies.

These insects have well-developed tracheal tubes and open spiracles, from which the water is excluded by the numerous hair-like projections. In *Dytiscus* the last abdominal spiracles are very large and are used during the bouts of surface breathing when the

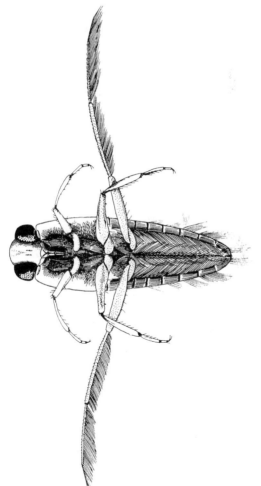

Fig. 15. Underside of water-boatman (*Notonecta*).

tip of the abdomen is protruded above the surface of the water. The other spiracles are chiefly serviceable in connection with the supply of air taken below in the special reservoir under the wing-cases. The store of air taken below is very useful in enabling the insect to stay down a much longer time.

This method of rising to the surface for supplies of air is quite a suitable method for these quick-moving, well-armoured insects. It is, however, very curious and interesting to find such elaborate contrivances in *Hydrophilus* and *Notonecta* for the passage of the air to the spiracles. See Fig. 15.

CHAPTER V

SURFACE-FILM BREATHING

IT is impossible to understand the breathing of a large number of aquatic insects, also it is impossible to understand the locomotion of many of them or the difficulties of exit from the water, without some knowledge of the properties of the surface-tension of water—the so-called surface-film.

WHAT IS THE SURFACE-FILM OF WATER?
Materials for each pupil.

Glass tumbler, needle, wire-gauze about a square inch, pipettes or fountain-pen fillers, a cork, some wire, screws, water.

56. Fill the tumbler nearly full of water and then add water gradually until the glass is overfull. Draw the shape of the surface with your eye level with the top of the glass.

57. Repeat the above with a light bead (imitation pearl) floating on the water. Explain the different positions it takes according as the glass is not quite full or overfull.

58. Dip a finger into the glass of water and lift it out carefully. You can lift up a large drop. Make a drawing to show the shape of the drop.

59. Hold a needle horizontally between finger and thumb and drop it upon the surface of the water. Describe what happens. Examine carefully the surface-film where it is in contact with the needle. Describe what you see.

60. Repeat above with a small square of wire-gauze instead of a needle.

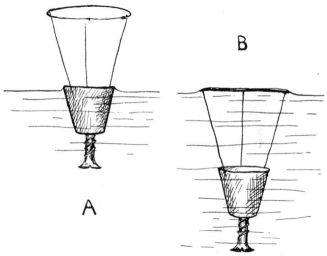

Fig. 16.   Mensbrugghe's float to illustrate the surface-tension of water.
    A.   The float in its normal position.
    B.   The float has been completely submerged. The "skin" prevents the float from rising to the position in A.

61. Make a Mensbrugghe's float. Look at Fig. 16. Be careful that the wire circle is horizontal and no sharp ends of wire sticking up, and that the screw is of sufficient weight to allow the cork to float with the top level with the surface, as in Fig. 16 A. Completely submerge the whole float by pushing down on top of the cork with a glass rod or pencil, and then release it. Explain why it takes up the new position.

62. Help the wire circle to break through the surface-film and explain carefully what happens.

These simple experiments show the existence of a certain physical condition in the water at the surface which is not shared by the rest of the water. We call it for convenience a "film," but it is in no way different from the rest of the water. It is probably due to the fact that the particles at the surface are in closer union with one another and with those just below than the particles of water elsewhere. There are invisible bonds of attraction binding particle to particle, and those at the surface not having water particles above them are drawn into rather closer contact with those alongside and immediately below. This part of the water will be denser, more difficult to break through and will act like a film, but a film which can instantly reform. The name **surface-tension** is a good one because it indicates one of the most important properties of this film. It is so tenacious that it prevents the water from overflowing (see Question 56); it defies the pull of gravity (see Question 58); it supports objects many times heavier than itself (see Questions 59 and 60).

It is surface-tension that rounds the rain-drop and the soap-bubble; it is surface-tension that causes liquids to rise in narrow tubes, and enables our bath sponges to hold so much water.

There are many most interesting experiments upon the surface-film of water and the part it plays in Nature, but they cannot be mentioned here. Those wishing for a more extended study of the subject are referred to *Object Lessons from Nature*, Vol. II, Miall (Cassell's); *Soap Bubbles* by Prof. Boys (S. P. C. K., 2/6); *Round the Year* (chapter on Duckweed), Miall (Macmillan, 3/6).

SURFACE-FILM BREATHING.

Material for each pupil.

Culex gnat larvæ, Dytiscus larvæ. Corked collecting tubes 3″ by 1″. Glass tanks. Lens. Microscopes. If possible larvæ of Dixa and Stratiomys for comparison, saucers of water. Slides of Dytiscus and Culex larvæ.

Fig. 17. Larva of *Dytiscus marginalis* breathing.
Photographed in a tank.

Dytiscus larvæ in glass tanks. If the larvæ are scarce, four pupils can work at one tank, two on each side.

63. How could you tell this was an insect ?
64. Briefly describe the chief features of the larva.
65. Make a drawing about life-size.

66. Is it lighter or heavier than water ?

67. What is its attitude when resting on the bottom of the tank ? Is there more than one attitude ?

68. Describe its attitude when rising to the surface.

69. Which part of the body is the first to reach the surface ?

70. How is the surface-film used ? Are there any special structures used in connection with this ?

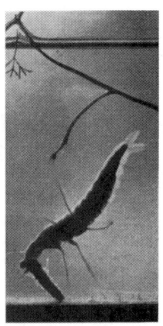

Fig. 18. The capture of food (caddis-worm) by *Dytiscus* larva.

71. Make drawings from the side and from above of the breathing attitude.

72. Where are the spiracles ? How many ?

73. Disturb the larva, and describe carefully what happens when it stops breathing and goes below.

74. Does the larva take any visible supply of air below ?

75. Examine, if possible, the hinder part of a larva under the microscope. What further information does this examination give you? Draw, and indicate the air-tubes in your drawing.

Although for convenience in the arrangement of this book, one problem only is dealt with here, in practice it is often advisable to extend the observations to include locomotion and feeding while the larvæ are available. It will be easy for the teacher to frame suitable questions to bring out the further interesting points in connection with the life of these creatures. See Miall, pp. 39—49.

Culex larvæ in corked, glass collecting tubes three parts filled with water. A dead specimen on a piece of white paper.

76. Examine carefully the living and the dead specimens, and write a general account of the larva (kind of thing, size, colour, shape etc.).

77. Draw it natural size, and then enlarged 4 times.

78. Is the larva heavier or lighter than water?

79. How does it get to the surface for air?

80. Which part of the body is the first to reach the surface?

81. Watch and describe carefully what happens at the surface. Take out the cork and watch again from above.

82. How is the supporting power of the film made use of? Compare with *Dytiscus* larva.

83. Bring a pencil point near to the larva to disturb it. How does it free itself from the film? Is it an effort to do so?

84. Examine the hinder part of a larva under the 1″ power of a microscope. Where are the spiracles? What prevents water entering them when submerged? Make a drawing of these parts. Compare your drawings with Figs. 19 and 20.

For further details read Miall, pp. 97—103.

If these larvæ are kept they will turn into the second

stage or pupæ, which are very active little tadpole-like
creatures (see Fig. 21). The breathing arrangements
of these pupæ are entirely different from those of the
larvæ, and takes place by means of two trumpets on
the back of the thorax. This change of the breathing

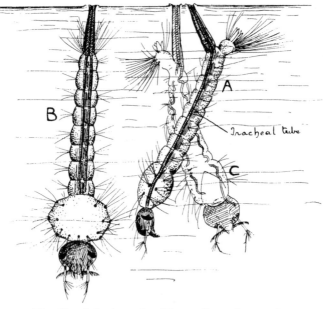

Fig. 19. *Culex* larvæ breathing at the surface. × 7.
A. Side view of larva. B. Dorsal view of full-grown larva.
C. Cast skin of a larva. Notice that the lining of the tracheal
tubes is left behind with the cast skin.

position should be carefully noted and a drawing made.
The reason for it will be clear later.

There are many insect larvæ that use the supporting
properties of the surface-film for breathing purposes,

and of course it must be remembered that water-beetles and water-boatmen use the surface-film in connection with their breathing. It is the surface-film which prevents the water from entering the spiracles, and which enables the insects to carry air below entangled among the hairs on the abdomen.

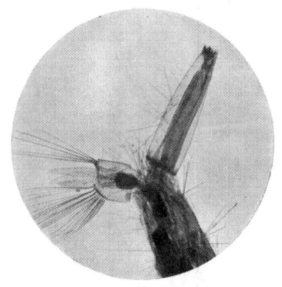

Fig. 20. Photomicrograph of the posterior end of a *Culex* larva, showing swimming tail and breathing tube.

If the interesting larvæ of *Dixa* and *Stratiomys* can be found or procured, and they are both much more common than is usually supposed, other fascinating uses of the surface-film can be studied. *Dixa* larvæ not only use the film for breathing purposes, but bent U-wise they can climb up the sides of the aquarium or up the stems and leaves of water plants dragging with them an encircling film of water. *Stratiomys* with its beautiful tail plume spread out on the surface or enclosing a bubble as it wriggles below fascinated two of the world's most famous naturalists, Réaumur and Swammerdam.

For further details, including Swammerdam's figures

Fig. 21. Photomicrograph of *Culex* pupa.
(Notice the breathing tubes.)

of *Stratiomys*, see Miall, pp. 189—193. For *Dixa* see Miall, pp. 157—164.

Thus, as well as the perfect insects which rise to the surface for their supplies of air, there are a number of insect larvæ which do the same. They show some special device for quickly piercing the surface-film, some method of clinging to the film during the breathing period and of quickly escaping from the hold of the film if danger threatens. Most of them are heavier than water and so hang clear from the film, and can sweep round in search of food at the same time.

The chief modification for breathing in the case of *Dytiscus* and *Culex* larvæ is the suppression of all spiracles except the last pair, which are carried at the extreme end of the abdomen or on a special breathing tube. Connected with these spiracles are a pair of large tracheal tubes which distribute the air throughout the body. See Figs. 17, 19 and 20.

CHAPTER VI

DIVING-TUBES AND A DIVING-BELL

Material for each pupil.

Ptychoptera larvæ, rat-tailed maggots (Eristalis), water-scorpions, water-spiders. Saucers with sand at the bottom. Glass tanks with water and weed in which water-spiders have been living for a day or two. Microscopes. Pocket lens. Black paper.

DIVING-TUBES.

Examine the *Ptychoptera* larva in a saucer of water with some sand at the bottom.

85. Describe the general appearance of this larva. A piece of black paper put underneath will show it up more clearly.

86. What is the head like?

87. Describe the tail.

88. Examine body with the lens. How many segments can you count? What about the surface of the body?

89. Make a drawing of larva life size and then magnified 4 times.

90. Watch the larva on the sand in the saucer of water. What happens to the larva and to the tail ?

91. Add water to that in the saucer or remove the larva to deeper water, and thus find out the limit of extension of the tail.

92. Examine the tail of your specimen under the low power of the microscope. How many tracheal tubes can you see ?

93. Where are the spiracles, and how many ?

94. Make at least 2 drawings to show the structure of the tail.

For further details see Miall, pp. 170—173.

The work with the rat-tailed maggot—the larva of one of the drone flies—is very similar; in fact Questions 85—94 should be repeated, and in the written descriptions resemblances and differences between this larva and *Ptychoptera* should be carefully noted.

This larva is most interesting not only on account of its long extensible tail but it has a wonderful filtering apparatus in its mouth, and when in its winged stage it so closely resembles a bee that most people would be deceived by it. It was because of this resemblance that the Greeks made the mistake of supposing that bees could be bred from a rotting carcase. A most interesting account of this is given in Miall, pp. 198—218, where there is also a full account of this larva and Réaumur's observations described.

Examine living and dead specimens of the water-scorpion (*Nepa*).

95. Describe the general appearance of the insect (size, shape, colour etc.).

96. Draw the insect life size, top view.

97. Examine, describe and draw the head with its piercing proboscis.

This kind of proboscis or trunk is both a piercing and a sucking apparatus, and is found in all members of this family (*Hemiptera* or Bugs). When convenient look at the proboscis of a water-boatman, green fly (*Aphis*), pond-skater.

98. Examine and describe the forelegs of this insect. Give it something to catch (a small worm is convenient). Of what does this catching leg remind you ?

99. Examine and describe the breathing tube. How does it differ from that of *Ptychoptera* ? Is it extensible ? Are there air-tubes inside it ?

100. Where are the spiracles ?

101. Watch the creature in the tank with a quantity of weed. How would it catch its prey ?

102. Write notes upon the use of the tube. Compare Fig. 22.

103. A dead specimen should be displayed on a piece of cork linoleum, with wings outstretched. This will help in getting a clear understanding of the tube.

104. Make one or two drawings to show the tube, its character and its attachment to the body.

The larvæ of the water-scorpion have a very short breathing tube. If they are available for examination, notes and drawings should be made.

It is comparatively easy for a rapid swimmer and a well-protected insect like a water-beetle to rise to the surface for its supplies of air, but it is a different matter for a sluggish insect or one not adapted to swimming.

Men when working below the surface of the water adopt some such device as that of *Ptychoptera* or the rat-tailed maggots, and have supplies of air sent down a flexible tube.

This curious diving-tube method has not been adopted by many insects, for it is adapted to very special cases. Both the larvæ of *Ptychoptera* and *Eristalis* live more or less buried in their source of food. Under these circumstances the need for rapid

Fig. 22. Water-scorpion (*Nepa*) breathing.
Photographed in a glass observation tank.

locomotion does not exist, true legs tend to disappear and a roughening of the body or the provision of short stumpy processes is adopted to give it the necessary purchase for movement in the food. The head tends to become reduced and mouth parts simplified. The breathing is undertaken by those spiracles which are more likely to be able to reach the air, the hindermost pair, and in order to do this more successfully the posterior part of the abdomen is formed into an extensible tail. Needless to say there is a limit to which the length of the tail can extend, and therefore the larvæ are to be found in the shallow water around the edges of muddy ponds.

The water-scorpion with his short and non-extensible breathing tube has a different problem to face. He is a bug; he lives upon the juices sucked from other animals, preferably living ones. In order to catch his prey, he crawls about the weed near the surface; his shape, colour and sluggish habits enable him to escape detection, and his breathing tube will be most valuable during the periods of waiting for the passing of some creature. Then the clasp-knife arrangement of the forelegs will catch its prey and draw it to the piercing proboscis.

A DIVING-BELL.

Although the series of lessons in this book is based upon the study of water insects, it may not be out of place to consider an animal which is closely allied to insects, yet not an insect.

Spiders are terrestrial, they are air-breathers, and just as the struggle for life has sent some insects into the pond, a spider has likewise changed its environment and adopted a water habitat. The problem of breathing has also been faced by this spider, and solved in a truly wonderful way.

The spider is commonly called water-spider (*Argyroneta aquatica*).

Examine water-spiders in glass tanks with weed. They should have been put into the tanks a day or two before the lesson. See that the tanks are covered with pieces of glass to keep the spiders from wandering.

105. Watch the spider in the water and then induce it to crawl out. Describe the differences in its appearance under the two conditions.

106. Write down a short description of the spider as seen out of the water, and point out why it is not classed under Insects.

107. Make one or two drawings to show its general appearance and, if you can, one to show its appearance when under water.

108. Examine the bubble-like nest or air-bell. How is it made?

Ask permission to look at the air-bells in the other tanks. In this way a number in various stages of construction will be seen, and it will be easier to answer the question.

109. Draw the air-bell in your tank and then interchange tanks with one or two others so as to draw different stages.

110. How does the spider get the air into the diving-bell?

111. In what ways does the principle of the surface-film help the spider? Or, in other words, why does the air remain in the air-bell, and how can the spider carry down so much air?

112. Name any other creatures that carry air below as a film on the outside of the body.

We have now seen the interesting way in which the water-spider provides for its breathing when under the water. The method reminds us of the diving-bell in which men work at the foundations of piers and break-waters. They are sent down in a huge iron bell into which air is pumped. The spider builds a mesh of threads among the weeds and having a very hairy body, it takes down a large amount of air entangled among these hairs. It kicks off some of the air by means of its legs and the bubble ascending is caught in the mesh of threads. Repeated visits to the surface and the repeated kicking off of air will add to the imprisoned bubble, and as it gets larger the mesh-work of threads is extended down the sides. Thus a stationary diving-bell is formed, closed above and at the sides, but open

Fig. 23. The water-spider (*Argyroneta aquatica*) returning to its diving-bell home.

below, and into this bell the spider can go at pleasure. It is used as a home, as a watch-tower from which to rush out on enemies or prey, and as a nest for holding the eggs, rearing and tending the young spiders until they are ready to go off to seek their own fortunes.

Water-spiders are quite easy to keep in captivity, and the rearing of young has been observed in an ordinary aquarium.

CHAPTER VII

A NEW INVENTION

(TRACHEAL-GILL BREATHING)

Material.

Dragonfly larvæ (the small demoiselle and one of the larger kinds). Dead specimen of demoiselle larva. Mayfly larvæ (any one of the common kinds). Caddis larvæ. Stonefly larvæ. Alderfly larvæ. Glass tanks. Saucers of water. Powdered carmine and pipettes, or a paint-box will do. Pocket lens. Microscope.

Examine the larvæ of the small demoiselle dragonfly (*Agrion*) in a saucer of water.

113. Write down reasons for calling it an insect.

114. Describe its general appearance (kind of thing, shape, colour, size, parts of body and appendages).

115. Make a drawing, natural size.

116. Examine a dead specimen of this larva and make a more careful examination of the head, thorax and abdomen. Describe any special features of each part.

117. If possible compare several larvæ, or get leave to look at the larvæ of your neighbours. Do you notice any differences in the plates on the thorax ? Why is this ?

118. Make a drawing, enlarged 4 times, of the end of abdomen with the tail plates.

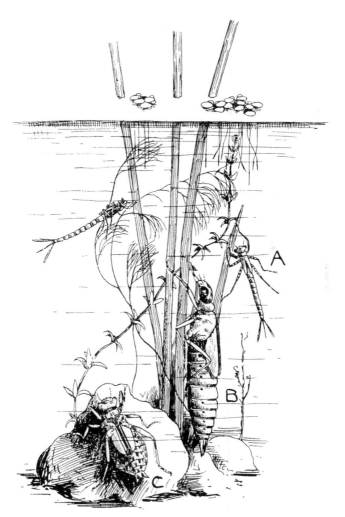

Fig. 24. The three common kinds of dragonfly larvæ. (Nat. size.)
A. *Agrion.*   B. *Æschna.*   C. *Libellula.*

It is often interesting and instructive to examine a living larva under the low power of a microscope. For very small larvæ a hollowed slide is often convenient, but this will not do for larvæ of mayflies, dragonflies, caddis and such like. An efficient and useful live box can be made from a glass microscopic slip, a little plasticine and a rectangular cover glass; if this last is not available, another slip will do quite well. The plasticine is rolled into a cylinder about 2″ long and of a diameter rather more than the width of space required between the two sides of the live box, say about $\frac{1}{4}''$ or $\frac{2}{10}''$. This roll of plasticine is then bent into a semicircle and placed upon the glass slip, and the rectangular cover or second slide placed upon the plasticine and pressed down evenly until a watertight joint is made. This live box will hold water, and can be used with an ordinary microscope or the lantern microscope.

119. Examine dragonfly larva in such a live box under the 1″ objective of the microscope. Are there any additions to your description of the larva that you can now make, or new discoveries to report?

120. Examine and describe the tail plates, and make a drawing of one plate under the 1″ objective.

121. What are the tubes that you see inside it?

These tail plates are a new kind of breathing organ which we have not met before. Each plate has within it a number of branching tracheal tubes which are connected with the main tracheal system of the larva. Such a structure is called a **Tracheal gill.** The leaf-like character of this gill will enable the air in the tracheal tubes within it to effect an exchange of gases, receiving oxygen from the water and discharging carbonic acid gas. A tracheal gill is not a blood gill such as we get in the external gills of the tadpole and the internal gills of fishes, for in a tracheal gill it is mainly the tracheal or air-tubes which are brought into close contact with the oxygen-containing water and not blood. See Figs. 25 and 27.

122. Is there any action which would make these plates more efficient as breathing organs?

123. Place some larvæ into water which has had its dissolved air removed. This is done by prolonged boiling, and then allowing it to cool. What happens? What is the result of this experiment?

124. Remove the head from a dead specimen, and extend the arm-like mask or catching device. Describe it.

125. Mount a mask in dilute glycerine and draw it under the low power.

126. Watch a living larva supplied with small worms or bloodworms. You may be able to see the use of the mask in capturing living prey.

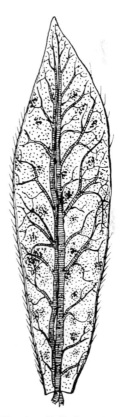

Fig. 25. Tail plate of a demoiselle (*Agrionidæ*) dragonfly larva.

The larvæ of the larger dragonflies (*Libellulidæ* and *Æschnidæ*), if obtainable, should be examined in glass tanks.

127. Repeat Questions 114, 115, 117.

128. Examine the end of the abdomen. How does it differ from that of the *Agrion* larva ? Make a drawing.

129. Watch the larva in the tank ; bring a pencil near to it to disturb it. Notice its method of getting through the water. How is it accomplished ?

130. Remove the larva into a saucer of water and again disturb it. Can you tell now how the sudden dart forwards is done ? (See Questions on locomotion.)

131. If it is still obscure use a little powdered carmine rubbed up in water, and with a pipette place some colour near the tip of the abdomen. A little vermilion or light red from the paint-box will do instead of carmine powder if the latter is not available. Describe fully what you see.

The hinder part of the food canal (the rectum) is enlarged to form a good sized chamber, the walls of which are well supplied with tracheal tubes. Water is drawn into this chamber through the anus, and then expelled by muscular action. If this action is a vigorous one the force of the expelled water propels the creature forwards. For details of this rectal chamber see Miall, pp. 338 and 339.

Besides these two methods of tracheal gill breathing, the one by means of tracheal tubes in an external plate and the other by means of tracheal tubes lining a rectal chamber, dragonfly larvæ have a more natural way of breathing. The larvæ of all dragonflies have thoracic spiracles, and these are functional during the later stages of the larval life. If larvæ of the *Libellulidæ* family are available, further interesting work can be undertaken as under.

132. Examine the thorax of one of these larvæ ; find the spiracles on the dorsal surface of the thorax between the segments. Make a drawing to show their position and size.

133. Place an old larva (one with wing plates well developed) in a beaker of water, and gently warm the beaker. Watch carefully these thoracic spiracles and note whether bubbles of air come from them. If so, what does it prove ?

134. Place young and old larvæ in water that has been previously boiled to get rid of the dissolved air. Put some reeds or sticks for the larvæ to climb up if they so desire. Describe carefully what happens. Do the young and old larvæ behave alike ?

135. What do you conclude from these experiments as to the relative value of the two methods of breathing in the young and old larvæ ?

Experiments of this kind were devised by Dewitz, and a short account of them with his conclusions is given in Miall, pp. 335—7.

There are quite a number of other aquatic larvæ that have adopted the tracheal gill method of breathing, and they will be briefly referred to below. It will not be necessary to add here questions to bring out the detailed structure of each of these larvæ. Such questions can be readily framed by the teacher or by the pupil if a fuller examination of each insect is thought advisable. The questions given refer mainly to the tracheal gills.

Examine some **mayfly** larvæ in a saucer of water. Any of the common kinds of mayfly will do.

136. Where are the tracheal gills ?

137. How many are there ? Describe their shape ?

138. Make two drawings—one to show the larva and the position of the tracheal gills, and the other a tracheal gill enlarged.

139. Examine a living specimen in a live box under the 1″ objective of the microscope. Describe the detailed structure of a gill.

140. Are there any movements of the larva or of the gills which will make the gills more efficient ? If so describe them, and explain the benefit of such movements.

For further questions on mayfly larvæ see Questions
211—216.

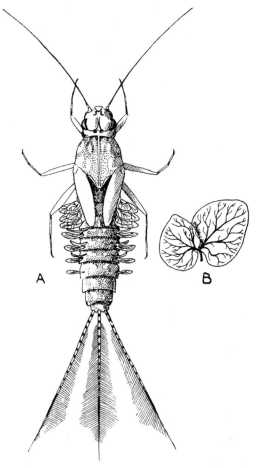

Fig. 26.   A.   Dorsal view of mayfly (*Chlœon*) larva.
B.   Tracheal gill of same.

Examine some **caddis larvæ** in a saucer of water.

141. Describe the case—its shape, materials, doors. Pull an empty case to pieces, and describe its construction.

142. Draw side view of case, and the front and back doors.

143. Remove a larva from its case. This is most easily done by gently inserting a pin through the back door. This will make the larva begin to crawl out of the case, and with a little encouragement from the pin point and a push from the pin head most larvæ can be dislodged. But **don't** try to pull it out.

Examine the caseless larva, and describe its general structure.

144. Make a drawing of larva from above. Colour it.

145. Has the colour any connection with living in a case? Watch one in its case before answering this.

146. Can you see anything along the sides of its body or on its back which you take to be tracheal gills? Describe what you see.

147. If you haven't represented these gills in your sketch, do so now.

148. How many gills are there, and how are they arranged?

149. Can they move? Is there any movement of the larva which will be useful in helping these gills to fulfil their purpose? Watch the larva which is out of its case.

150. Remembering that the larva lives in a case, are there any special disadvantages in connection with its breathing?

151. How does it get over them?

152. Test the flow of water through the case of another specimen by means of powdered carmine or a little paint. Place a little coloured fluid first at one end of the case and then at the other. Which way does the water flow? Why does it flow? Describe the experiment and what it proves.

153. Can the larva keep a current of water flowing through the case when its head and legs are out of the front door working at case building or hunting for food? This will require careful observations to answer.

154. What will keep the body steady so that it can be thrown into waves and thereby keep a current of water flowing through the case?

To answer this difficult question completely would require much time and much careful observation. It was solved in a very ingenious way by giving a caddis larva little pieces of thin glass or mica with which to build a case ; for if you turn a larva out of his case, he will at once begin to build another, and by restricting the materials you can get him to build of almost anything. You might try larvæ in different saucers with beads, tea leaves, pieces of cork, india-rubber, chips of wood, paper. If you want to have your larva build a glass case which will let you see what he is doing, you must choose the larva called *Phryganea grandis* which builds a case with the leaves of a water plant carefully arranged like tile work. See Fig. 82 (1) in Miall. If you succeed in this, you will then be able to answer this question. You will also get some help by another examination of the larva. Direct your attention to the first abdominal segment, the one immediately after the segment with the brown spots.

155. How many processes can you see ?

156. Describe their shape and exact position.

157. Are they contractile ?

158. Make an enlarged drawing of this part of the larva to show these structures.

159. Examine very carefully the last abdominal segment. Describe any curious structures, and make a drawing to show these things.

160. Can you tell now why it is useless to try to pull a larva out of its case ?

With the knowledge got from these further observations and inquiries, you should be able to give a fairly complete answer to Question 154. In connection with this read about Mr Taylor's experiments with caddises in Miall, pp. 247—9.

161. Examine a small larva in a live box[1] with the 1″ objective under the microscope. It will be well to glance quickly at the whole animal, moving the box about so as to see various parts of the larva. In this way you can check your previous observations. Then examine more carefully the hair-like tracheal gills. Describe their appearance and make a drawing to show this and the tracheal tubes inside. Compare with the gills of the mayfly.

Examine **alderfly larvæ** (*Sialis*) in saucers of water.

162. Write a description of the larva, being careful to mention all the various parts of the body, especially the tracheal gills.

163. Make a drawing, life size, naming all the parts.

164. Compare the tracheal gills of this larva with those of the caddis and mayfly, in respect to number, position, shape etc.

165. Is there any "breathing" action?

Examine **stonefly larvæ** (*Perla*) in saucers of water.

166. Write a description of the larva.

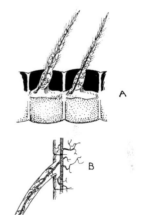

Fig. 27. Tracheal gills.

A′. Side view of the tracheal gills of the alderfly (*Sialis*) larva.

B. Diagram of a tracheal gill showing the connection of its tracheal tubes with the lateral trunk tube.

167. How would you distinguish it from a mayfly larva?

168. Make a drawing (enlarged 2—3 times).

169. Can you find any tracheal gills? What about the tufts at the bases of the legs?

[1] See p. 44 for description of a home-made article.

170. Examine a larva under the microscope in order to see whether these tufts have tracheal tubes in them or no. If they have, then they are tracheal gills.

171. Compare these tracheal gills with those of the mayfly, alderfly and caddis.

172. Is there any "breathing" movement?

This method of breathing by means of tracheal gills which has been called a new invention is unique as far as I know. A large number of aquatic insects find it impossible to rise to the surface for their supplies of air; a mayfly or stonefly larva clinging to the under surface of a rock in a rapid stream, an alderfly larva crawling about the mud or a caddis imprisoned in its case do not want to face the risks of rising to the surface. It is obvious also that the tracheal system must be closed (in dragonflies spiracles are used as well), and that fresh air must be obtained from that dissolved in the water.

To do this well is the problem that has been so successfully solved by these insects with this tracheal gill method of breathing. It is simply some method of bringing the air, which is inside the tracheal tubes of the insect, into intimate contact with the water and its dissolved air. Branches from the main air-trunks of the body pass into thin plates or filaments which are in such a position as to stand out freely into the water. There is almost always some body movement, or movement of the gills themselves, to keep the water about the gills in motion and so bring fresh supplies of aërated water to the gills. This movement is of utmost importance to case-dwelling larvæ, for the water inside the case would soon become foul. We have noticed how the caddis keeps a steady current of water passing through the case.

This method of breathing is very interesting, for it is a compromise. The tracheal or air-tubes of the insects are used, and yet the creature remains submerged. It seems to be able to get sufficient pure air from the water by means of a principle called diffusion, which is the name given to the passage of gases or liquids through thin membranes, which in this case will be the thin skin covering the gill plates or filaments.

Another interesting feature is the great variety of form of tracheal gills. Surely this is another proof of our early statement that aquatic insects have descended from terrestrial forms and have invaded the water. The problems of breathing have been solved independently, and even in this one method of submerged breathing the great variety of solutions shows that each one has adopted a form best suited to its own life.

All these insects which breathe by means of tracheal gills are aquatic during part only of their lives. It may be a very large part, but at last they leave the water, acquire wings and breathe like normal insects.

173. Write a general account of the tracheal-gill breathing among aquatic insects.

There is yet another method of breathing used by aquatic insects, not a new invention, for a large number of aquatic animals, from worms to tadpoles, use it, but yet distinctly unusual for an insect to adopt. It is easily observed in the larvæ of the harlequin-fly (*Chironomus*), to which larvæ the common name of **bloodworms** has been given. This method of breathing is by means of **blood gills.**

To save repetition the questions upon this method have been included among the others in Chapter X, where the life-story of this insect is fully dealt with.

Of course, to make a more complete survey of the breathing problems, the consideration of this method should come here, and it will be quite easy to turn forward and answer the questions upon the bloodworm's breathing. Then when Chapter X is done in ordinary course, it will only be necessary to review briefly this problem of its life.

Several of the smaller worm-like larvæ, which appear in the aquarium, have no visible means of breathing ; some have traces of tracheal tubes, mere vestiges which proclaim their terrestrial ancestry, but which cannot be used ; none shows any trace of tracheal gills. They must have oxygen. How do they get it ? From a knowledge of the breathing of other small aquatic animals we conclude that the larvæ can get the necessary oxygen from the water by diffusion through the skin of the whole body, instead of having a special structure like a gill for this to take place. Their small size makes this method sufficient for their needs.

Reviewing the whole subject of the breathing of aquatic insects there is one point that requires emphasis. Although the larvæ of aquatic insects may breathe through spiracles, or by means of tracheal gills, or blood gills or merely the skin, the perfect insects, or imagos as they are called, always breathe gaseous air by means of spiracles and tracheal tubes. Some of these perfect insects (beetles, water-boatmen, water-scorpions) retain their aquatic habit even in adult life but, unlike the larvæ, they are free at any time to leave the water and fly away if the need arises.

CHAPTER VIII

LOCOMOTION

I. How a land insect walks.

Material.

Living and dead cockroaches[1]. *Sheets of white paper.   Petri dishes.   Pocket lens.*

174. Examine a dead cockroach. Turn it over on its back, and carefully remove the three legs of one side. This will require great care, and should be done with a penknife and the legs not pulled off with your fingers or you will leave behind the large flat first joint attached to the thorax. Lay the legs in order on a piece of white paper.

175. Examine the second leg. At a first glance, how many divisions do you see ?

176. Examine more carefully with the lens, and count the number of separate pieces or segments of the leg.

The names given to the parts of an insect's leg are as follows.   Beginning with the first part at the end attached to the thorax: (1) coxa; (2) trochanter, a very small triangular piece between (1) and (3); (3) femur; (4) tibia, often very spiny ; (5) tarsus consisting of a number of small segments, the last one of which carries the claws.

177. Describe the coxa, femur, tibia and tarsus of the leg of the cockroach.

178. Make a careful drawing of the leg, magnified 4 times. Name all the parts.

[1] Any large running insect would do, such as the ground beetle (*Carabus* family), or, better still, the bloody-nose beetle (*Timarcha tenebricosa*).

179. On the under side of the tarsus are some white pads. How many are there ? What use are they put to ?

180. Compare the three legs. Note resemblances and differences.

Examine living cockroaches. I have found for the questions which follow, a piece of white paper and half of a Petri dish the best way of watching the living insects. Whilst the cockroach is getting used to the new surroundings the following questions can be done.

181. What is the position of the 6 legs when the insect is at rest ?

182. Make a few quick sketches, natural size, to show the insect's body and the position of the 6 legs.

Fig. 28. Cockroach walking.

In drawing such a thing as a living insect or indeed any small living animal that will not keep still for long, it is best to have a good look at the creature, commit the shape and proportions to memory, and then make a faint complete sketch. This can be improved by further observation and criticism. Don't be afraid to leave your sketches rough and unfinished ; several rapid impressions are often very valuable when it is not easy to make a more elaborate drawing.

When the insect is quieter and more used to the surroundings, the inverted half of the Petri dish can be carefully raised and the insect allowed to wander. Keep the glass cover ready to drop over the insect if it is too energetic or is likely to escape on to the floor. Pupils that make slow and quiet movements will succeed best with these and all living creatures.

183. Watch and describe the method of using a leg. How much of the leg touches the ground and forms the "foot"?

184. The amount of contact of the leg with the ground can be shown by making a cockroach walk over smoked paper or over wet printer's ink on to white paper.

185. Watch and describe the movements of all 6 legs as they work together. How many legs are always in contact with the ground? How do the first and third pair of legs differ in their method of working? What is the use of the middle pair of legs?

It would be quite useful to extend this general inquiry so as to include the use made by the legs for other things than locomotion. For instance, the use made of the legs for cleaning purposes is easy to observe. The common house-fly continually uses its legs to clean wings and body, and then the legs themselves are cleaned. If a living cockroach is dusted over its head and antennæ with flour or starch dust, the cleansing of these parts will be easily seen. The forelegs of ants, bees and wasps, if observed under a low power of the microscope, will be seen to have a beautiful comb for the antennæ. For information upon this subject see *Natural History of Common Animals* (Latter), *Round the Year*, and *Useful and Injurious Insects* (Miall).

II. SWIMMERS.

A. **Water-beetles.**

Material for each pupil. *Glass tanks with large water-beetle. (Dytiscus or any member of the Dytiscus family.) Dead beetle*[1]*. Lens. Microscope. Small rubber bands.*

It is best to have a little weed in the tanks: water starwort (Callitriche) will be most suitable, so that the insects can cling to it for periods of rest.

186. What are the locomotory organs ?

187. Do all the legs share in the work ?

188. Investigate and describe the work done by each pair of legs. This is best done by fastening the legs by rubber bands so that they cannot be used and then carefully watching the effect. Various experiments should be tried. First fasten 2 pairs of legs so as to see the work of 1 pair only at a time. Then legs on one side can be fastened, and 1 pair of legs in turn put out of use.

189. Are the legs in any way specially adapted for swimming ?

190. Do the legs of each side work together or alternately ?

191. Watch the swimming stroke and describe it. Is it at all like rowing with an oar ? Is there anything like "feathering" ?

192. Has the shape of the body any part to play in locomotion ? Describe this fully.

193. Examine a dead specimen, move the legs and investigate their attachment to the thorax. Describe what you find, and give· reasons for the change.

194. Make a drawing of the under side of the beetle to show the way the legs are attached. There is no need to draw the whole of each leg.

195. Examine the third pair of legs with a lens or under a low power (2 inch) of the microscope. Describe carefully all the parts, comparing with the typical insect's leg of the cockroach. Draw the leg, enlarged.

[1] One or two dead specimens could be made to serve a class of ordinary size.

196. Can you get any idea of the rate of locomotion ? A stop watch will be necessary.

197. When the beetle is at rest below, which legs are used to hold the beetle ?

198. When the beetle is breathing at the surface, what part do the legs play in this ?

B. **Water-boatmen.**

Material for each pupil. *Glass tanks with weeds*

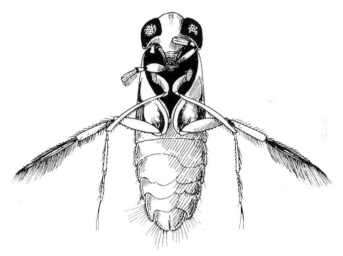

Fig. 29. Underside of water-boatman (*Corixa*).

as before, one of each kind of water-boatmen. Dead specimens. Lens. Microscope. Rubber bands.

199. Which legs are the oars of the boatman ?

200. Do the other legs take any part in locomotion ? Experiment with fastening the legs by means of rubber bands as in Question 188.

201. How are the legs adapted for swimming ? Why is the term "oar" a good one ?

202. Do the oars work together or alternately ?

203. Are the oars feathered ? Describe the stroke from start to finish.

204. Has the shape of the body anything to do with its power of swimming ? Compare carefully the differences between *Notonecta* and *Corixa*.

205. Which is the better swimmer ? Why ?

206. Compare the swimming attitude of each. Has this any-thing to do with the breathing attitude ?

207. Draw a dead specimen of each, enlarged 4 times, to show shape of body and the position of the legs.

208. Examine under a lens and then under low power (2 inch) of the microscope, the oars of the two water-boatmen. Describe and draw.

209. Can you get any idea of the rate of locomotion ?

210. When at rest below, which legs are used to hold the insect?

III. THREE MAYFLY LARVÆ.

Material for each pupil. *Living and dead larvæ of mayflies of Bætis, Ephemera vulgata, and Chlœon species. 3 saucers of water. In one a piece of rock, in another some sand. Lens.*

For comparison. Dragonfly (demoiselle) larvæ.

211. Write down the things in which these 3 larvæ are alike.

212. Write down the main differences in size, colour, shape etc.

213. Watch *Chlœon* in the saucer of clear water, *Bætis* in the one containing the rock, and *Ephemera vulgata* in the one con-taining the sand. Describe the locomotion of each and the natural habits of each.

214. Watch all three kinds swimming in a saucer of water. Which is the best swimmer ? How is the swimming accomplished ? Do they all swim in the same way ?

215. Examine the tail filaments by means of a lens, and make a drawing to show them and their structure.

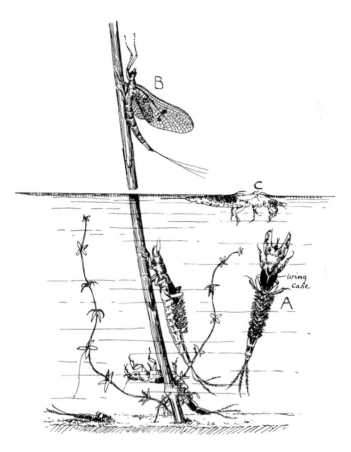

Fig. 30. Mayflies (*Ephemera vulgata*). Four larvæ, one
burrowing in the sand. (About nat. size.)

A. An old larva or nymph with wing cases very prominent.
B. Mayfly just emerged from the skin C, left floating at
the surface.

216. Make a thorough examination of a dead specimen of each larva; examine the shape of the body; examine the legs, compare and make drawings; examine the gills. Write a careful account. From what you have now discovered and from what you found out in Question 213 you should be able to describe how each kind of larva is adapted to the life that it leads, and show the connection between their structure and the place where they are found. Illustrate your account by several drawings.

In these three mayfly larvæ we have an excellent chance of discussing the way in which structure and habit of life go hand in hand. These three larvæ are very different, and yet they are all obviously mayfly larvæ.

The free-swimming larva of *Chlœon* (Fig. 26) may be taken as the normal form and comparison made with *Ephemera vulgata* (Fig. 30), which has gone in for burrowing, with a consequent change in legs and tracheal gills, and with *Bœtis* which lives clinging to the underside of rocks in a swiftly flowing stream. The flattened form of *Bœtis*, legs adapted for clinging and plate-like gills are all again special changes connected with its curious life.

Although locomotion is the chief thing in our thoughts, we shall be wise to take a broader outlook and, when a chance such as this offers, discuss other related problems as well. The thought of the evolution of these two special kinds from some more normal ancestor will help us to realise a similar diverging evolution among higher animals.

217. Compare the swimming of the larvæ of demoiselle dragonfly.

IV. CRAWLERS.

Material for each pupil. *Sialis* (*Alderfly*), *caddis and Ptychoptera larvæ. Saucers of water. White paper. Small pieces of black paper. Sand. Lens.*

Alderfly larvæ in saucers of water.

218. Describe its method of locomotion.

219. Bring pencil point near to the creature's head. Describe what happens.

220. Can it swim ? Drop it into deeper water.

221. Remove the larva from the water on to a piece of white paper. You need not be afraid that the larva will die, even if kept out some time, unless it gets very dry. I have known them keep alive for days in damp air. Watch how it gets along on the paper. Describe its movements, and compare carefully with those in the water.

222. Compare the movements of the legs with those of the legs of the cockroach.

Examine **caddis larvæ** in the same way.

223. Describe the locomotion of the caddis. Explain exactly what happens to the case.

224. Can it swim ? You cannot give a complete answer, unless you examine all kinds and all ages of caddises. You can answer for the kind of larva under examination.

225. How does the larva draw its case about with it ?

226. Have the legs of a caddis different movements from those of the cockroach ?

227. Remove a caddis from its case, and compare its locomotion with and without the case.

228. Make a drawing of caddis crawling with case on its body.

Examine **Ptychoptera larvæ** in saucers of water with a little sand at the bottom.

229. Watch the larva in motion. Describe how it gets along. What other animals get along in a similar way ?

230. Are there any legs ? Any pro-legs (that is stump-legs, like those on the abdomen of a caterpillar) ? A piece of black paper put underneath the insect will help you in these questions.

231. Are there any other structures which would help loco-motion ?

232. Make a drawing of the larva (enlarged 4 times) showing all these structures clearly.

233. This larva is called a maggot. Can you explain why the *Ptychoptera* maggot, like the blowfly and bee maggots, has no true legs ?

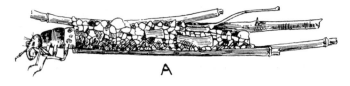

A

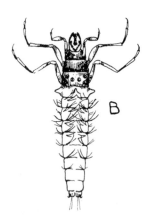

B

Fig. 31. A. Caddis larva crawling with its case. × 2.
B. Dorsal view of caddis larva taken out of its case. × 2.

V. Wrigglers.

Material for each pupil. *Bloodworms (Chironomus larvæ) in saucers of water. Culex larvæ in glass tanks of water. Sand. Lens. Microscope.*

234. Examine the bloodworms in the saucers of water. Watch their methods of locomotion. The ordinary one is a crawl. Describe how this is done. The addition of a little sand to form a rougher surface will make this clearer.

235. Are there any legs or pro-legs which are useful for this purpose ? Examine with a lens and then under the low power (1 in. or 2 in.) of the microscope. Describe carefully what you see, and make two or three drawings (enlarged) to show any interesting structures.

236. Make the larva swim. This is usually done by disturbing it. Watch very carefully, and explain its method. Make a diagram to explain the movements of the body.

237. Can this larva direct its course in any way ?

238. Compare with *Culex* larvæ in the tank. This is a very lively larva. Watch its method of swimming, and carefully describe it.

239. Are there any fins or other swimming organs ? How does it differ from the bloodworm in its method of wriggling ?

240. Examine a larva under the low power (1 in.) of the microscope. Describe and draw any structure which would help it to get along better when wriggling.

241. Which is the better swimmer, *Culex* or the bloodworm ?

242. Can you give reasons why one should swim better than the other ?

243. Can *Culex* direct its course through the water ? If so, how can it do so ?

244. Visit a pond and bring home all kinds of larvæ. Describe and draw any that get through the water by wriggling. If you cannot name them nor find their names in any of your books, keep them in saucers or small dishes covered up with pieces of glass, and see what they turn into.

VI. Skaters.

Material for each pupil. *Pond skaters and whirligig beetles. Saucers of water. Dead specimens as well will be very useful. Lens. Microscope.*

Pond skaters.

245. Revise the lesson on the surface-film (pp. 26—28).

246. Watch the pond skater. Describe its locomotion.

247. Are the legs used in the ordinary insect (cockroach) way?

248. Describe the surface-film at the place where the feet rest.

249. Why do not the feet break the film? To answer this question it will be necessary to examine the feet of this insect under the microscope. A dead specimen will do, or slides of the legs could be made previously ready for use. Describe fully and make drawings.

250. Remove the pond skater to a piece of paper on the desk. Describe carefully its locomotion on land and compare with its locomotion on water.

251. What happens if you frighten the insect? Does it dive below? Have you ever seen one below the surface? Submerge one and describe what happens.

252. Describe the attitude of rest.

253. Make drawings (enlarged) to show its attitude when on the surface-film.

Whirligig beetles.

254. Watch and describe the movements of these beetles. If you have seen them on a pond or stream, describe what you saw then.

255. Try to trace out the path taken by one beetle. Make a diagram to show this.

256. How does this movement differ from that of the pond skater? What is in contact with the film in each case? How do they compare as to speed?

257. How are the legs adapted for this skating movement? To answer this examine a dead specimen. Remove the legs of one side and mount them in order, either temporarily or permanently, and examine under the 1 in. power of the microscope. Write a careful account of these legs, with an enlarged drawing of each, and explain the special points about each.

258. How does the beetle behave if frightened?

259. Place some on the surface of a glass tank of water and make them go below. Describe what happens, and the method of locomotion when submerged.

260. Make a drawing of the beetle (× 4) from above.

Further interesting points which can be referred to are : the curious antennæ, the divided eye for vision in air and water at same time. See Miall, pp. 30—35.

Fig. 32. Larva of a dragonfly (*Æschna*) swimming across the tank. Nat. size.

VII. Special methods.

Material for each pupil. *Dragonfly larvæ (either of the two larger kinds—Libellulidæ or Æschnidæ). Phantom larvæ. Simulium larvæ. Saucers of water. Glass tank. Powdered carmine or a paint-box. Pocket lens. Microscope.*

Dragonfly larvæ.

261. Watch the slow stealthy crawl of the larva. (This of course is not a special method.) But disturb the larva by means of a pencil, and describe what happens.

262. A little powdered carmine or vermilion paint mixed with water should be put near the tip of the abdomen. A fountain-pen filler makes a convenient pipette for this purpose. Patience and repeated trials may be needed. Compare Questions 129—131. Describe the result of this experiment and explain fully the method of locomotion.

This curious and interesting method of locomotion is not without its parallel in another group of animals. The octopus and cuttlefish adopt a somewhat similar device, for they also take water into a special chamber for breathing purposes, and then this water, if expelled forcibly, will drive the creatures forward at a considerable speed. It would be interesting and instructive to get a specimen[1] of the

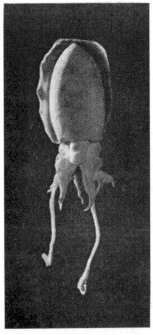

Fig. 33. Cuttlefish or *Sepia*. ×¼. Notice the funnel projecting from the mantle fold. It is through this funnel that water used for breathing purposes is expelled, and, if expelled forcibly, it is a means of locomotion.

[1] The Laboratory, Marine Biological Association, Plymouth. 1s. each.

common cuttle (*Sepia*), and, by cutting open the mantle (respiratory) chamber, show this arrangement in the animal. See Fig. 34.

Fig. 34. *Sepia* with mantle cavity exposed. Notice the two gills with the rectum between them and the funnel which has been opened by a longitudinal cut.

Phantom larvæ.

263. Examine phantom larvæ in the glass tanks of water, using a pocket lens. Why phantom ?

264. Write a brief description of the larva.

265. Draw it in its position of rest, side-view, ×4.

266. Watch carefully for any movement. Does it move slowly, or how?

267. Time its periods of rest between the sudden jerks. Take a number of records, and then find average.

268. How is it able to keep still so long, and how does it alter its position?

269. Examine a living larva in a hollowed slide or in a live-box. Are there any legs or organs of locomotion? Examine the larva carefully, and describe all that you can find out about it. Examine carefully the air-sac floats.

270. Make drawings of the larva to show its air sacs and swimming fin.

Simulium larvæ.

271. Examine *Simulium* larvæ in saucers of water. Briefly describe the general appearance of the larvæ, and make a drawing (natural size) to show this.

272. What is the method of locomotion? What other animals does it remind you of?

273. Put a piece of black paper in the saucer and induce the larvæ to creep on to it. Do you learn anything more from this?

Simulium[1] larvæ are found in fast-flowing streams, some kinds in rocky and some in weedy streams. They all require well-aerated water. They live attached to the stones or leaves of water weeds and creep about.

274. Test their holding on power by holding your saucer at an angle under the tap and allowing at first a gentle stream and gradually a stronger rush of water to flow against them. Describe the experiment. What conclusions do you come to?

275. Repeat the above experiment, but dislodge one or two larvæ whilst the water is flowing gently past. Watch carefully what happens.

[1] To keep these in captivity for more than a few hours will require running water.

276. Suddenly lift up a piece of weed or paper with a few larvæ on it, out of the water. Probably one or more will let go. Describe very carefully what happens.

277. Do you know any other animals that can immediately spin a silken thread and drop out of danger, and then climb up again ? Explain what advantage this device will be to the *Simulium* larva living in a rapid stream.

Fig. 35. A favourite stream near Ackworth School where mayfly, stonefly and *Simulium* larvæ abound.

278. Examine a living or dead larva under the low power of the microscope. Or, better still, slides of the larva can be prepared beforehand. Make a thorough examination of the larva. Write notes on its structure, especially of its head, pro-legs and sucker. Make one or two enlarged drawings.

For further details see Miall, pp. 175—182.

We have now taken a rapid survey of some of the modes of locomotion found among the aquatic insects, and it is instructive to compare the various ways of solving the problem. Locomotion is easier in water than on land, because the water carries most or all of the weight of the body, and this means that muscular actions can almost entirely be devoted to propulsion. If a general review of the whole animal kingdom is taken, it will be seen that the simplest animals are all aquatic; it requires quite a complex organisation of the body to walk on land, and still more to rise into the air. Insects in their evolution from some primitive aquatic ancestor in the very far distant past have conquered the land and the air, and it is not surprising that these most adaptable creatures have been able, when circumstances have driven some back into the water to live, to reconquer this medium also. Aquatic insects have gone from the land into the water, and we can easily see that there must be locomotory problems to be solved in connection with this change of medium, just as there are the problems of breathing. The equilibrium and mode of propulsion are very different in water from on land. The fact also that there have been several distinct invasions of the water means that there will be many different ways of solving the problem, and that the solutions here must bear a close relationship to the solutions of the breathing problems. A rapid and effective method of locomotion is essential if frequent visits to the surface for air are necessary, but a sluggish creeping habit will do for an insect breathing beneath the surface by tracheal gills.

The study of the modes of locomotion is a very useful exercise in adaptations, and through all our inquiries should run the query, "What connection has this with the insect's life?" or simply the query "Why?" An answer cannot always be given, but the attitude of mind which puts such a question is a most important thing to cultivate.

Let us summarise the results of this chapter. Some larvæ have gone back to a primitive method of locomotion found among the worms. *Ceratopogon* (Miall, Fig. 46), without any legs, swims by serpentine undulations of its long thin body, and *Chironomus* (bloodworm), a larva of a stouter build, forms a "figure of eight" by looping the body first one way and then the other. *Culex* and phantom (*Corethra*) larvæ have the addition of tail fins for striking the water, and the latter larva, which is almost transparent, has wonderful air-sac floats in its body for enabling it to remain in a horizontal position awaiting its prey of waterfleas and other small creatures. At intervals, by a quick lash of its tail, it suddenly changes its position. The larvæ of *Ptychoptera* and *Simulium* are also without true legs. *Ptychoptera*, after the manner of a maggot, has roughened segments as well as hook-bearing pro-legs; it advances through the mud by contractions and expansions of its body, the roughened surface giving the necessary purchase. *Simulium* with its powerful suckers holds fast in the rush of water or creeps, like a leech, about the rock or meshwork of threads, ready to cast off at the end of a new thread if danger threatens, and hauls itself back again when the danger is past.

Insects with true legs can crawl, cling, burrow or swim. Some, like the mayfly, stonefly and one of the dragonfly larva, swim by lashing the water with a tail of feathered filaments or flat plates, their legs meanwhile pressed against the sides of the body. Others, with legs fringed with hairs, row or paddle with great effect. The water-boatmen and water-beetles represent this group : indeed the *Dytiscus* beetle, with its rigid smooth body and even contour, with its powerful hind legs, is a wonderful swimmer. Some dragonfly larvæ can suddenly dart forwards by the expulsion of water from the rectal chamber ; and, lastly, some insects can walk, jump or skate upon the surface-film as though it were a sheet of ice. The pond skater, with specially prepared feet, glides over the surface without breaking the film, and the whirligig beetles disport themselves on the surface, darting hither and thither by the powerful strokes of their flattened legs.

For further information on the modes of locomotion of insects, see the various chapters in Miall, especially the summary in Chap. XIII. If possible, read Letters XXII and XXIII (Motions of Insects) in Kirby and Spence's *Introduction to Entomology*[1].

CHAPTER IX

THE ESCAPE FROM THE WATER

ALL aquatic insects can, and the majority do, leave the water at some period of their life. We have already noticed that all aquatic insects breathe by means of open spiracles and tracheal tubes during their final life-period, the **imago** or winged stage, and it is in order to enter this air-breathing and winged stage that they leave the water. The problem before us here is the

---

[1] A most valuable book, which should be on the shelves of every lover of insects. Second-hand copies are frequently on sale.

change from one medium to another, and the dangers and difficulties of the change. The reasons for this change will be discussed in the next chapter.

It is not easy to plan definite class-work for the consideration of this problem, because we cannot control the times of change nor order the insects to emerge when it would be suitable for us. However, by keeping a number of aquaria in the schoolroom, it is possible to do something at the subject, but it will require freedom to interrupt a Mathematics or Geography lesson in order to witness the escape of some insect whose compulsory change has not been timed to occur during the Nature Study lesson.

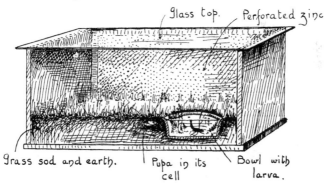

Fig. 36.   Drawing of a box arranged for studying the pupation changes
of *Dytiscus* and other aquatic beetles.
The front part of box and bowl has been cut away to show the
arrangement.

I. CRAWLING OUT.

Material. *Full-grown larvæ of Dytiscus, of the alderfly, and of dragonflies in suitable aquaria.*

It is quite possible to get the full-grown larva of the *Dytiscus* to crawl out of the aquarium and change into the pupa in an earthy cell. Arrange your aquarium, which should be a shallow dish, like a pie-dish, to imitate a pond. This can be done by having a grassy sod all around, and by means of sand, rocks and weed an easy passage out of the pond can be made. The *Dytiscus* larva will then be able to crawl out of the water, make a cell in the earth, moult his larval skin, become a white beetle-like pupa, gradually darken, and finally awaken as a beetle. Unlike most aquatic insects, the *Dytiscus* beetle goes back again into the water, but he retains the power of crawling out and flying to another pond. If your aquarium is not covered, you will most likely lose your beetles, for they often leave the water and fly at night time.

279. If you have a full-grown larva, fit up an artificial pond. Describe and draw your "pond," and write full notes with dates of all the things you can see. If you are very careful, it is possible to open the earthen cell and to have a movable lid so that you can watch the changes within. Make drawings to illustrate your notes.

Read Lyonnet's account of his experiments in Miall, pp. 69—71.

The alderfly larvæ need very similar treatment, but they often walk considerable distances before pupating. The larvæ crawl out of the water and pupate in the earth.

280. Keep some alderfly larvæ in a similar pond to that in use for *Dytiscus*. Write notes, and make drawings of the escape and subsequent events.

Dragonfly larvæ, or nymphs as they are sometimes called, when ready to emerge, require reeds or sticks up which to climb. They will not require any earth round the aquarium. If you see any larvæ escaping from the water, try to answer the following questions.

281. What is the position of the insect ?

282. Draw or photograph it.

283. How long before anything happens ?

284. Describe exactly where the skin splits.

285. What comes out first ?

286. Take the times of the various changes, and describe the whole process.

287. What is left behind ?

Fig. 37 A. The escape from the water. Dragonfly (*Libellula*) just out of its nymph-skin.

288. Describe the size and colour of the newly-emerged dragonfly. Do these change ? When ?

289. Measure the wings when first emerged, and again when fully expanded. What interval of time between these two ? What is the power used to expand the wings ?

290. How long before it can fly away ?

291.   Make drawings or take photographs of the various stages.

292.   Read Tennyson's description of this change in *The Two Voices* (verses 3—5).   Do you consider this an accurate description ?

II. ESCAPE AT THE SURFACE.

Material.   *Living mayfly nymphs.   Caddis pupæ. Culex and Chironomus pupæ in suitable aquaria. Lens.   Carmine.*

Fig. 37 B.   The escape from the water.   Dragonfly (*Libellula*) drying its wings.

The large mayfly, *Ephemera vulgata*, is the best one for observation.   The nymph or full-grown larva can be told by the prominent wing-cases on the back of the thorax.   They can be observed in a shallow pie-dish or glass tank. It will be best to have a few reeds or rushes sticking out above the water for the newly-emerged mayfly to cling to.

293. Watch and, if you see the emergence, note very carefully the time taken and the details, as much as you can see in the time.

294. Does the skin split in the same place as that of the dragonfly?

295. Examine an empty skin. There will be some left on the surface by mayflies escaping when you were not watching, or you can collect any number from your mayfly stream. Find out exactly where the split takes place, and examine the empty skin under the microscope (1 in. objective) and write notes upon your observations. Notice carefully the tracheal gills.

296. How does the mayfly prevent itself from drowning?

297. Can the mayfly fly away at once?

298. Examine the newly-emerged mayfly with a lens. Notice the transparent shroud-like skin completely covering it.

Fig. 38. The empty skin left behind when the dragonfly (*Æschna*) escapes from the water.

299. Make some drawings to illustrate your notes and to show in picture form the method of escape.

The newly-emerged mayfly is covered with a transparent skin which is usually moulted after a period of rest on a reed or twig.

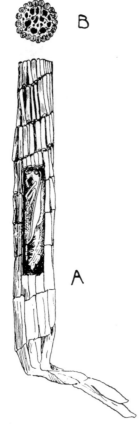

Fig. 39. *A*. Pupa of caddis (*Phryganea grandis*) in its case. A part of the wall of the case has been cut away to show the pupa inside.

B. Front door of case showing silken grating.

But many mayflies do not cast this skin or only partially, and they often mate and lay their eggs without assuming their final form. The full beauty of the wings and tail filaments is not seen unless this last change of skin takes place.

Caddis and its difficult escape. Collect a number of caddis cases which have a silken grating over the front door. In some cases small stones also partially block the entrance.

Also collect empty caddis cases for examination.

300. Look up your notes and drawings on Questions 141—161.

301. Describe and draw a case to show the method of closing it. Why is this done?

302. Is there anything inside the case? Can you find out without opening it?

303. Experiment with powdered carmine or paint with some of these cases in a saucer of water. What do you conclude from this?

304. Carefully take a case to pieces. Is there anything in the nature of a cocoon?

305. Examine the inhabitant. Compare with the larva which you have already seen.

This creature is the **caddis pupa**, and resembles a sleeping moth.

306. What moth-like characters do you see ?

307. Make a side-view drawing, × 2.

308. Watch the living pupa in the saucer of water. Is it able to move ? Describe any movements and explain, if you can, their use.

309. How does the pupa breathe ?

310. How does this pupa differ from the pupa or chrysalis of a true moth ? If you don't know what a moth or butterfly chrysalis is like, ask your teacher to show you one.

311. The caddis fly, a moth-like fly, lives in the air in the neighbourhood of ponds and streams. How can the insect make this double escape, firstly from the case and secondly from the water ? How does the soft and feeble pupa or fly get out of the case-prison ?

This question is a difficult one to answer fully unless you are fortunate enough to witness the emergence, either in captivity or by the side of some pond or stream. It is much less uncommon than most think, and I have seen it fairly frequently. However, a good deal of information can be obtained from the pupæ and from empty cases.

312. Test the toughness of the silken grating. You will appreciate the problem of the escape.

313. Examine some empty caddis cases out of which the insect has escaped. The remains of the silken grating will give you a clue. Look inside. Do you find the empty pupal skin ? Would you say that the pupa or the fly had escaped ?

314. Have you ever seen the empty pupal skin of the caddis ? If so, where ? Next time you visit the ponds and streams where caddises live, look out for them.

315. Examine the pupa. Write careful notes upon the head and all that you can see. Make a drawing.

U.   6

316. Compare your drawing and notes with Fig. 83 D and p. 260 in Miall.

317. Can you now explain the whole process of the escape from the case?

318. Describe, if you can, the second escape from the water. The position of empty pupal skins will furnish you with important evidence.

The caddis pupa is furnished with special apparatus for tearing away the door of its prison. It is then free to climb out, and to climb or swim to the surface. The importance of having a very mobile abdomen, and legs and head not glued down like a chrysalis, is thus clear. The emergence into the air is sometimes like that of the mayfly, but more often the pupa, using its legs and abdomen, climbs on to a floating leaf or up the stem of a water-plant into the air and then divests itself of the pupal skin, which you can find empty and dry.

Culex pupæ in a glass tank.

319. Compare *Culex* pupæ with the larvæ (see Questions 76—84). Or, if the living larvæ are available, it would be interesting to have both larvæ and pupæ in the same tank. See Figs. 19 and 21.

320. Describe carefully the breathing attitude of each.

321. Make a drawing of the pupa at the surface, life size and then × 4.

322. Why is the equilibrium of the body changed, so that the breathing tubes are in a different part of the body? It is well to put this question to yourself, but the answer may be deferred or you can give your present ideas and then confirm or change them later.

323. Can you make out any fly characters in the pupa? Describe what you can see.

324. Watch the emergence of the gnat. Where does the pupal skin split?

325. What is the exact position of the pupa when the splitting takes place?

326.   What comes out first ?

327.   How long before the gnat is out ?

328.   What use does the gnat make of the pupal skin in getting out ?

329.   Examine and draw an empty pupal skin.

Fig. 40. The escape from the water. Photomicrograph of a male *Chironomus* fly which has failed to get its legs free from the pupal sheaths. The specimen was mounted as a microscopic slide.

330.   Is the gnat able to fly away at once ?

331.   Examine and compare the escape of the harmless gnat (*Chironomus*). See Chap. X, Questions 371—373.

332.   Describe the whole process of the escape. Can you now explain the change in the position of the breathing tubes (Question 322) ?

These insects described above are the ones best adapted for observation as class-work; there are many others with more fascinating methods of escape which although not so easy yet could be observed by a little care and patience. There are some too that do not make this escape, but live as larvæ and imagos in the water. However, these too can escape if they wish to do so, and take to their wings for courting purposes or with the desire to seek a fresh pond. The water-boatmen and water-scorpions are examples of this group. The pond skaters cannot properly be said to have an escape from the water, for they live on the water and probably enter it merely to hibernate in the winter time.

The escape from the water is a comparatively easy thing for large and vigorous larvæ able to climb out at the side or by means of water plants, but for others it is an event of considerable danger. There is the danger during the rise to the surface, and many mayfly larvæ and caddis pupæ fall as prey to the ever-watchful fish. There is always the possibility of mishap at the crucial moment, especially for those emerging at the surface, and many a *Culex* and *Chironomus* perish by a sudden capsize or by the sticking fast of a leg. Then there are the dangers and difficulties of the surface-film, only serious to small insects, but these difficulties are often turned into helps by the ever adaptable and in-genious insect.

To show how adaptable and ingenious they are, it will be well to consider one other remarkable escape.

SIMULIUM'S WONDERFUL METHOD.

Material for each pupil. *Simulium larvæ and pupæ in saucers of water. Pocket lens. Some arrangement to keep the pupæ under observation in running water will be necessary to witness the escape.*

333. Revise the structure and life of the larva. See Questions 271—278.

334. Examine and describe the pupa and its cocoon.

335. Examine the cocoon more carefully; pull one to pieces; describe its construction and material. Make some drawings to show their shape and attachment to rock or weed.

336. Remove a pupa from its cocoon. Is this easily done? Examine the pupa carefully and write notes upon it, pointing out the " fly " characters.

337. Of what use are the two tufts of filaments ?

338. Make a drawing of the pupa in its cocoon, × 2, and then one of a pupa removed from its cocoon, × 4.

339. Can you tell now why the pupa was difficult to pull out of the case ?

340. Investigate the cocoons, and find out whether the pupa leaves the cocoon to rise to the surface like the caddis or whether the empty pupal skin is left behind in the cocoon.

341. Where does the pupal skin split ?

Simulium larvæ live in rapid streams. We have already seen how the larva guards itself from the dangers of the rushing water by means of suckers and anchoring threads; but what about the pupa during its period of inaction and sleep ? To keep it from being carried away by the stream, the larva, previous to its change into a pupa, builds a cocoon of silk attached to the rock or stems of water-weeds. The change takes place in the cocoon and then, bursting the top of it open, the pupa protrudes head and shoulders with the breathing tufts into the rushing water. The roughened edges of the abdominal rings hold the pupa safely in the cocoon. Thus the pupa is safely anchored and the breathing tufts collect air for the developing fly.

Now comes the problem. *Simulium* flies are small and delicate, and you will find them in numbers on the bushes and trees near the stream: the pupa skin is left behind in the cocoon; how can this delicate fly pass from the cocoon, through the rushing water, negotiate the surface-film and enter the air?

342. Keep some pupæ in a vessel of running water. Both the larvæ and pupæ need a good supply of well-aerated water. Watch carefully, and try to solve this problem.

The method of escape is an unusual one. The breathing filaments gather air from the water. More air is obtained than is needed for the respiration of the pupa, and it is passed out into the space around the fly underneath the pupal skin. When this skin splits, a bubble of air comes out, and this quickly rises to the surface and bursts. When it bursts, a small hairy fly is shot out safe and sound on to the surface-film, over which it scrambles until it reaches some solid object on which it rests until able to fly away. The principle of the surface-film is used here to great advantage, for the ·fly arrives at the surface quite unwetted by its passage through the water. See Miall, pp. 175—188.

CHAPTER X

LIFE STORIES. FROM EGG TO IMAGO

IN the foregoing chapters three great problems of aquatic insects have been dealt with—the problems of breathing, locomotion and escape from the water. There are many other problems facing aquatic insects— the capture of food, the escape from enemies, the problems of reproduction, mate-hunting and egg-laying, and the problem of the distribution of a species over as wide an area as possible. These problems are not peculiar to aquatic insects but are shared by all other animals, so do not call for special mention here ; yet the problems of distribution and reproduction are very important, for they furnish a clue towards understanding the question of **Insect Transformations**—such a characteristic feature of the life of most.

In this chapter the life story of a very common insect will be dealt with, so as to give an opportunity of discussing the various problems throughout the life-story of an aquatic insect, and this will also give an opportunity to discuss the question of the Transformations of Insects.

LIFE STORY OF THE HARLEQUIN-FLY (*Chironomus*).

Prepare a number of shallow dishes, pie-dishes or saucers, with water, mud and decaying leaf-debris. Capture a large number of bloodworms and put them into the dishes. It is best, as suggested on page 7, to cover these dishes with pieces of sheet-glass. These dishes of bloodworms should be started a week or so before they will be wanted so that some pupæ will be ready for examination and all the stages available without waiting.

Fig. 41.   Mud tubes of *Chironomus* larvæ.

I. THE LARVA.

It is convenient to start with this stage, and come round to it again.

Material for each pupil. *A small saucer of water containing pond sediment or sandy mud, with one or two bloodworms. This must be prepared the day before*

*and allowed to remain undisturbed. Freshly killed
larvæ. Needle. Pocket lens. Microscope.*

343. Examine the saucer with its sediment of mud. Is it evenly
spread over the bottom? Describe and draw what you see.

344. Move the mound with a needle. Does it break up easily?
What do you argue from this?

345. Gently press the needle at one end of the case and travel
along gently pressing repeatedly. Describe what happens.

346. What is the mound? Examine the others. Are they open
at both ends? Can you perform Question 345 either way? Have
you any idea how the mound is made? What is the best name
for it?

347. Examine the animal. Why is it called a "bloodworm"?
Are there any drawbacks to this name?

348. Do you know any other animals which are called "worms"
which do not belong to the worm division of the Animal Kingdom?

349. Make a thorough examination of the larva by means of
the dead specimen, as well as the living ones. Use pocket lens and
also the microscope. Write a careful description of the larva with
three drawings. Whole larva × 4. Head and tail ends much enlarged.

350. Watch and describe its methods of locomotion. See
Questions 234—237.

351. Of what does the red colour of the larva remind you?

The red colour of vertebrate blood is due to a
substance called Hæmoglobin, which has a wonderful
power of taking up oxygen and keeping it until it can
distribute it to different parts of the body.

352. Is the red colour of the bloodworm due to hæmoglobin
or to some other substance? To answer this question we must have
some test for hæmoglobin which we can apply. The test is by means
of spectrum analysis.

**Method of demonstrating spectrum analysis to
a class. Apparatus required.** *Lantern. Prism or
a carbon bisulphide prism-bottle. Tubes of blood,*

chlorophyll, didymium nitrate. Some method of making a narrow slit to go in front of the condenser of the lantern[1].

The lantern is used as the source of light which we call white light. Place a piece of card, with a vertical slit in it, in the slide-carrier. This arrangement will throw a narrow beam of light in a straight line with the lantern. Place the prism-bottle in the path of this beam. Two things happen: the beam of light is bent out of the straight and it is split up into the various coloured rays spreading fanwise from the prism. Instead of a narrow beam of white light, we get a wide spectrum showing the rainbow colours—red, orange, yellow, green, blue, violet. The lantern and prism can be turned so that the spectrum falls on a white screen. A piece of white paper pinned up will do. The next thing to show is the way in which different substances placed in the path of light absorb different rays and form what are called dark lines or absorption bands in the spectrum. Place a tube of didymium nitrate[2] in the path of the light before it passes through the prism. You will see characteristic dark lines where the substance has absorbed and stopped certain rays. The spectrum of any substance can thus be mapped out, and no two substances have a spectrum consisting of the same combination of lines. Thus a sure method exists of detecting the presence of substances.

The spectrum of blood can now be shown and

[1] If the lantern is one of the "Stroud and Rendall" type, it will be best to remove the front lens and fix a piece of wood, with a narrow slit, in front of the condenser. The slit is best made by two strips of metal bordering a hole (1″ by $\frac{1}{2}$″) in the wood. By fixing one strip and having the other adjustable various widths of slit can be tried.

[2] Use pieces of coloured glass if this is not obtainable.

carefully noted. Also show the spectrum of chlorophyll—the green colouring matter of plants. This can be got by chopping up some grass, killing it by boiling in water, pouring off the water and adding alcohol, and gently heating. The alcohol will dissolve out some of the chlorophyll. If the class takes up Botany, especially the way in which plants feed, this spectrum of chlorophyll will have an interest.

In this way the blood colour of the *Chironomus* larva can be examined. It is impossible to collect enough blood for analysis in the way just explained, but spectroscopes are made which can detect the smallest quantities of substances, and it would be easy to show that the *Chironomus* larva had hæmoglobin in its blood.

A very convenient spectroscope for use in classwork, if the time taken for each one to use it is not inconvenient, is the direct vision spectroscope. With this the complete spectrum is got by directing it like a telescope to the window, and the absorption bands can be seen if the substances are placed just in front of the slit, so that the light has to pass through them before entering the spectroscope.

353. Write a full account of the spectrum and the method of spectrum analysis.

354. Hæmoglobin has great power of storing up oxygen. From where does the bloodworm get its oxygen?

355. How does the oxygen get into the blood? Are there any gills?

356. Examine more carefully the last few segments of the larva. Describe the position and number of the blood gills or thin walled tubes. Make a drawing (enlarged) to show them.

357. Why are they in this position? The tadpole has external gills on its neck. Defer this, if you do not know, until a little later.

358. What special advantage is it to the bloodworm to have hæmoglobin? Where does it live? Is there any connection between these two questions?

359. Compare the earthworm. The blood of this animal also contains hæmoglobin. Is there any similarity in life to cause this? When does the earthworm do most of its breathing? Go out with a lantern into the garden some warm damp evening. Describe what you see.

360. Visit if possible a stream or tank which abounds in bloodworms during the day and again at night. Describe where the bloodworms are at each time.

If this is not possible, read Miall, pp. 129—131.

361. Boil some water in a small flask to drive all the air out of it. Cork it and allow to cool. Then remove the cork and allow carbonic acid gas to occupy the space above the water. Introduce 2 or 3 bloodworms, cork the flask again, and note what happens. How long can the larvæ live on the oxygen already stored up in their blood? You might ask your teacher to do this experiment for you. There is no need for each one to do it. Describe the experiment and the results.

362. Has the bloodworm any means of keeping a constant flow of water passing through the case or burrow? Compare the caddis in this respect.

363. Examine a young living larva under the 1 inch objective of the microscope. The larva can be put into a hollowed slide with a large coverglass over it, or in a small live box made as suggested on page 44. Then the coverglass can be pressed down gently until it presses on the larva and restrains its movements somewhat.

Review the general structure, and make out and describe the tracheal vestiges at the anterior end; the heart on the dorsal side just above the blood gills (Can you now answer Question 357?); the alimentary canal and salivary glands.

For details see Miall, pp. 122—129.

364. Find a large and dark coloured bloodworm. This will be nearing the end of its larval existence. What are the chief differences between this and a young larva?

365. Examine the saucers or dishes containing the stock of bloodworms. Can you see anything which might be the next or pupal stage ?

366. What proof have you that these are the pupæ of the blood-worms ? The answer to this is not clear yet, but it is well to have it in our minds so that we can remember to look for connecting proofs.

The store of larvæ will have begun to change into pupæ and maybe into flies as well. Capture all the flies that emerge and try to keep them alive in test-tubes plugged with cotton-wool. Or if they die, they may be kept for examination by preserving them in tubes of alcohol (70 %). Collect flies from near the stream or from walls and fences near by.

It is best to study the fly next, because it is difficult to understand the pupa and the whole problem of the transformation unless the chief features of the imaginal stage are understood.

II. THE IMAGO.

Material for each pupil. *Tubes containing living flies. Dead flies. Pocket lens. Microscope. Slides of the flies can be prepared beforehand.*

367. Examine a living fly. Describe its general appearance, shape, size, colour.

368. Describe and draw its posture when at rest. Do you see any reason for its name, harlequin-fly ?

369. Examine the fly more carefully by means of dead specimens and the microscope. Write a description of the fly.

370. What is the difference between male and female ? Draw the head of each (enlarged).

The fly is a very delicate insect with one pair of wings, and one pair of rudimentary wings or halteres. It thus belongs to the order **Diptera** (two-winged flies). The head is small and holds the large compound eyes, antennæ, and imperfect and useless mouth parts. The flies never feed. The easiest way of telling male and female is by the antennæ. They are plume-like and large in the male, and small and simpler in the female. The middle segment of the thorax is very large ; this is doubtless connected with the storage of muscles, for the only functional wings are carried by this segment.

III. THE PUPA.

Material for each pupil. *Chironomus pupæ. Empty pupal skins from the top of the water. If possible some pupæ ready to come out as flies. Have the stock saucers and dishes available for observation purposes. Pocket lens. Microscope.*

371. From what do the flies come out ? Have you any proof ? Has anyone seen it ?

372. Search the stock saucers and see if there is an example of failure to get clear out of the pupal skin. If so examine it, and compare it with Fig. 40.

373. If a fly emerges during the lesson, notice the following points : position of pupa, the split and its exact position, the order of coming out, time taken, condition of the fly when emerged, what it does, and then write an account of the whole affair. Compare Questions 324—331.

374. Examine a cast pupal skin on the surface of the water. Is it full of water or air ? How does the surface-film aid or hinder the process of emergence ?

375. Examine and describe the skin in more detail. What parts can you make out ?

376. Examine the pupa. In what respects is it like the larva ? In what respects is it like the fly ?

377. What is the position of the pupæ in the stock saucers? Why is this?

378. Is there any movement? Describe it and the reason.

379. What are the tufts on the thorax?

380. Make an enlarged drawing of the pupa to show as much of the structure as possible.

381. Examine the tail of the pupa very carefully. Can you see something like a transparent tube clinging to it? If you can see nothing, perhaps the pupa of your neighbours has this structure present, for it is not found on every pupa. If you see what is meant, examine it with a lens and try to find out what it is. Sometimes the cast pupal skins show these skins still clinging about their tail ends.

Fig. 42. Pupa of *Chironomus* with the larval skin (out of which it has just come) still clinging to it.

This observation, if made, furnishes sure proof of the passage or transformation of one stage into the next and, put together with the finding of a dead fly half emerged from the pupal skin, completes the succession from larva to imago. It is often possible to watch these changes, but it requires time and patience, and the first-named requirement is not easy to get in the already overfull curriculum.

The larva when full-grown, easily recognised by the thickened thorax, splits its skin behind the head and by means of blood pressure the head, eyes, antennæ, legs, etc. of the fly, which had been formed as it were inside out, are blown out and the pupa takes on a form not unlike the fly but with all the body covered with

the pupal skin. The empty and almost colourless larval
skin is pushed back and worked off, but in many cases
it clings about the end of the pupa as tell-tale evidence.
The pupa, although still red in colour, obtains oxygen
by means of two tufts of filaments on the thorax, and
these are slowly waved to and fro by the bending of the
pupa. In a day or two the pupa has gathered sufficient

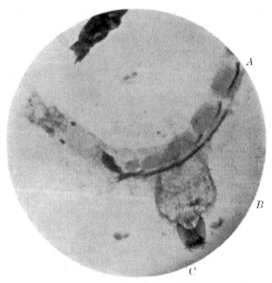

Fig. 43.   Photomicrograph of larval skin of *Chironomus* clinging to
the abdomen of the pupa.   *A*, Abdomen of pupa.   *B*, Skin of
larval thorax split open.   *C*, Larval head.

oxygen to make it buoyant and it floats up to the
surface, where it quickly goes through the last change
of skin. A crack appears along the back of the thorax
and the fly emerges into the air, using the empty skin
as a raft for support. Before long it flies away.

The life story is not yet complete. The flies have

to mate, and the females lay their eggs.   The mating can be observed in the summer, when swarms of these flies hover over the streams.   In a book dealing with the life and structure of this insect[1] some interesting details are mentioned.   The flies dance from 10—15 ft. from the ground, and occasionally a pair (male and

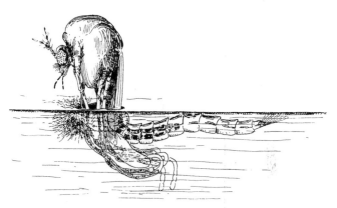

Fig. 44.   Female *Chironomus* fly emerging from the pupal skin.

female) leave the swarm and descend.   The female goes off to lay her eggs, and the male rejoins the dancing, singing swarm.   In one sweep of the net Mr Taylor, by whom these observations were made, found 700 flies, all of which were males.   This was on a calm evening, when pairing was easy.   On a windy evening in a swarm of 4300 flies, 22 were females.   It would be interesting to capture some of these swarms and confirm these observations.

[1] *The Harlequin-fly*, Miall and Hammond (Clarendon Press).

IV.   THE EGG.

Material for each pupil.   *An egg-mass in a saucer of water.   A piece of black paper.   Pocket lens and microscope.*

The gelatinous egg-masses can be collected in any number from the edges of water tanks or troughs.   A knife-blade is useful to detach the egg-masses which are moored to the side.

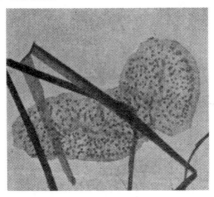

Fig. 45.   *Caddis* spawn.

382.   A piece of black paper underneath will show up the structure.   Describe what you can see (shape, size, colour).   Use a lens.

383.   The masses were found moored to the side of the tank. What moors them ?

384.   Is the mooring rope stretchable ?

385.   Dip one mass into hot water for a few seconds, and examine the attachment through the microscope.   Draw and describe.

386.   Examine the eggs in the egg-mass under the microscope, and describe their arrangement.

387.   Make 2 or 3 drawings, one life-size and others enlarged, to show the structure.

388.   For what purpose is the jelly mass ?

389.   Name other creatures which lay their eggs in jelly or mucilage.

390.   Why is the egg-mass moored by this peculiar interlacing rope ?

391.   Keep some of the egg-masses until the eggs hatch out. How do the newly hatched larvæ differ from the bloodworms you have already seen ?

392.   How long is it before (i) the newly hatched larvæ are like ordinary bloodworms, (ii) they are full-grown larvæ ?

For a discussion upon these questions and further details, see Miall, pp. 146—152.

Culex, the common grey gnat, will afford another convenient insect for the study of the life story. It is very common and it is easy to keep in captivity, and during the summer all the stages from egg-raft to fly can be got at the same time so as to have all the stages at hand for examination.

If more convenient, the life story of *Culex* could be taken in detail instead of that of *Chironomus*. See Miall, pp. 97—113.

Dragonfly larvæ, caddises and mayflies are easily kept in captivity, and the different habits and transformations are important for comparison with the life story of *Chironomus*. But they are not convenient for a general study—the life story is too protracted. Certain interesting details of the life story of these insects have been dealt with already in the foregoing pages.

There are a few problems suggested by our work in connection with the life of aquatic insects, especially by the work of the last chapter and this one. To put them into question form, they would be as under:

393.   Why do insects change their skins as they grow ?

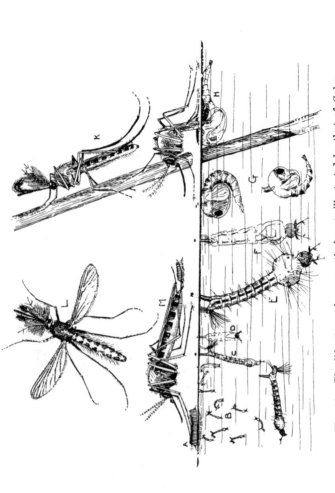

Fig. 46. The life story of an aquatic insect, illustrated by that of *Culex*.

A. Egg-raft floating on the surface. B. Young larva just hatched. C. Larva four days old. D. Cast skin of young larva. E. Full grown larva (2 weeks old). F. Cast skin of larva out of which a pupa has just come. G. Pupa—one breathing at the surface. H. Escape of female fly from the pupal skin after 4 or 5 days as a pupa. K. Side view of male fly—notice the carriage of the hind legs—this is a quick way of telling *Culex* from *Chironomus*, the other gnat, for *Chironomus* holds up the front legs. L. Dorsal view of male fly. M. Female fly laying her eggs.

394. Why do most insects have well-marked stages in their life history (larva, pupa, imago) ?

395. Why do some insects have a resting stage (pupa) ?

396. Why do most insects acquire wings ?

397. Why is there often a change of mouth parts and change of food as well ?

These are far-reaching questions and do not admit of simple answers, yet they should be considered. The life of all insects is governed by this wonderful power of transformation, and to aquatic insects the changes are especially important.

The body of insects is covered with chitin, a secretion from the true skin (see Chap. II), and this makes a profound change in the method of growth. The chitin cannot grow, so that when the covering is too small, a larger new one is formed underneath and the old one is cast. These periods of silent growth and the abrupt visible change make it possible for the insect to grow new structures, which may only become outwardly visible after the change of skin.

A few insects only grow bigger at each moult, and never change their shape or acquire new structures ; some others gradually acquire wings; whereas others, and these the greater number, change very strikingly and have their life divided into very distinct stages. The **larval stage** first, which bears no resemblance to a highly specialised **winged stage** called **imago**, and between these two usually occurs a resting **pupal stage.** The larval stage of many of the insects of this third group is very like, in its shape and structure, certain lower or more primitive wingless, adult insects ; in the case of others, the abundance of food and the consequent easy life have brought about a degeneration, and the larvæ become still simpler. The subsequent changes of skin provide a means of replacing lost organs, and of forming new and more elaborate ones for use in the winged stage. These changes are seen more clearly after the last but one and the last moult. The contrast between larva and imago is often so marked that if the connection between them was not known, they would be placed in widely different classes of insects.

The larva is always fully grown before these important changes take place. In this fact lies its peculiarity, for it is what is called an **adult** transformation. In the transformations of many other animals, for instance in the crab, the changes occur at a different time. The minute creatures hatched from crab eggs pass rapidly through a number of striking changes which they complete *before* they are full-grown : in the insects, the changes occur *after* they are full-grown. The transformations of the frog give us another example of adult transformation; for the tadpole, in its structure and life, approaches very closely certain primitive adult ancestors of the frog. Indeed both the insects and the amphibians give us, in different ways and from different causes, an abridged summary of their ancestral history in their own life stories. They reclimb their own ancestral trees.

Every insect, indeed every animal, has to feed and grow, to separate from its fellows, seek for mates and find an uncrowded site. In many animals the distribution is done while the animal is very small, especially if the adult is sluggish. In the crab the young stages live in the floating life of the sea, and can be carried immense distances by tides and currents. The adult stage is sedentary and, when the transformations are over, the tiny crab settles down among the rocks where it feeds, grows and in time becomes sexually mature.

In insects, on the other hand, the feeding is done first, and when the full size is attained the questions of reproduction and distribution are considered. Wings are usually developed, and in many insects changes in the mouth parts are also made. The acquisition of wings is a means of seeking uncrowded sites for egg-laying.

Wings can be acquired without much disturbance of the life of the insect, as for instance in the cockroach, where they get gradually larger at each moult. In the mayflies and dragonflies the same thing occurs, but the aquatic habit causes a more abrupt appearance of the wings after the escape from the water. In this group the body of larva and imago are very similar in shape and size, and the mouth parts are also similar. But to take full advantage of the use of wings, the imago becomes very active and the larva, relieved of all questions of dispersal, becomes very sluggish. With the great activity of the imago goes the need for a lighter body, a richer and less cumbersome food, and the necessary changes in muscles and internal parts, and, in some cases, a great change of mouth parts. The most complex insects (save the ants) have learnt to sip the sweet juices of flowers. The wonderful double evolution of a nectar-sipping insect (with all its wonderful appliances) and an entomophilous flower (with its colour, shape, scent and nectar) from simpler ancestors, cannot be discussed here.

Complex changes from a more or less sluggish, voracious larva to a light, dainty, winged imago with the necessary change of mouth parts, need a resting pupal stage. To take a simile; it will be impossible to change your pattern of fire-grate unless there is a period without a fire. Thus a period of rest from eating is necessary if the pattern of mouth parts is to be changed. In short, change of food means a resting pupa. The pupal stage in many land insects is so arranged as to coincide with the difficult winter season. In most aquatic insects the winter is passed in the larval stage, and I have frequently obtained a large selection of insect larva even when snow is on the ground.

Insects are classified into three groups according to their transformation, and the classification of aquatic insects is given below.

I. **No transformation** (ametabolic).

No change in form, no resting stage.

   Podura.   Miall, Fig. 111.

II. **Partial transformation** (hemimetabolic).

Almost all acquire wings, which become visible externally as plates on the back in the early stages. No marked change of food, mouth parts, form of body, length of legs. No resting stage.

   Mayflies (*Ephemeridæ*).   Miall, Chap. VIII.
   Dragonflies (*Odonata*).   „   „   IX.
   Stoneflies (*Plecoptera*).   „   „   VII.
   Bugs : (water-boatmen, water-scorpion, pond-
        skaters and water-measurers) (*Hemiptera*).
        Miall, Chap. X.

III. **Complete transformation** (holometabolic).

There is always a resting pupal stage in which the insect never feeds. Wings are developed internally, and do not show externally until the larva changes into the pupa. Usually a striking contrast between larva and imago in form of body, mouth parts and food.

Alderflies (*Neuroptera*). Miall, Chap. VI.
Caddises (*Trichoptera*).   „   „   V.
Gnats and other two-winged flies (*Diptera*). Miall, Chap. II.
Beetles (*Coleoptera*). Miall, Chap. I.

Quite a number of insects and many among the aquatic forms (*Chironomus*, mayfly, stonefly, alderfly), never feed in the winged stage. The question of the feeding of the imago is connected with the time which elapses between the escape from the water and the processes of mating and egg laying, and also with the manner and place of the latter act. If mating and egg laying is a simple business and there is no time lost in seeking mates, as in the mayflies and *Chironomus*, the life of the winged stage may be very short.

For a further discussion of this subject, see a paper in *Nature* by Prof. Miall[1].

[1] "The Transformations of Insects," *Nature*, Vol. LIII, p. 153, Dec. 19, 1895.

APPENDIX I

MATERIAL

Dealers who supply **living aquatic insects** are :

Mr Thomas Bolton, 25, Balsall Heath Road, Edgbaston, Birmingham.

Mr L. Haig, Beam Brook, Newdigate, Surrey.

Mr J. E. Molloy, Bell Cottages, Cowley, Middlesex.

It is impossible to give any exact idea of prices for these will fluctuate according to the time of year and the difficulty of capture. But speaking generally, the summer is the best term for this study, and as an indication of prices I will quote a few from the list of Mr Haig.

Beetles.

		s.	d.	
Dytiscus			9	per pair
Hydrophilus		1	6	,, ,,
Hydrobius			3	each
Acilius			4	,,
Whirligig (*Gyrinus*)			3	,,
Water-scorpion (*Nepa*)			4	,,
Water-boatman (*Notonecta*)			4	,,
Water-spiders (6)		1	6	

Larvæ.

		s.	d.	
Dytiscus		2	6	per doz.
Caddis			6	,, ,,
Sialis			6	,, ,,
Dragonfly		2	6	,, ,,
Bloodworms (*Chironomus*)			6	a box of 100.

Teachers unable to capture their own supplies or needing to supplement their own finds should write to these dealers for price lists. Mr T. Bolton has had many years' experience in supplying biological material to Colleges and Schools and is able to supply all the aquatic insects mentioned in this book.

Other material required for some lessons :

Caterpillars (living) from any well-known dealer such as L. W. Newman, F.E.S., Bexley, Kent.

Centipedes, found under large stones and logs. They should be handled with care.

Cockroaches (living) from Mr T. Bolton, 25, Balsall Heath Road, Birmingham.

Shrimps (boiled) from the fishmonger.

Spiders, collected from the garden or from ledges and dark corners especially in dark sheds, and preserved ready for use.

Wasps and other perfect insects can be collected as occasions offer and preserved.

KILLING AND PRESERVING OF INSECTS.

For observations into details of structure, dead, preserved insects are most convenient. The teacher should begin to get together a small collection of aquatic and other insects which are likely to be of use in classwork.

KILLING.

1. Chloroform and ether. Very useful for beetles, cockroaches, dragonflies, flies, wasps. These require care in handling and must be kept under lock and key.

2. Hot water. This is by far the best insect-killing agent. Death is instantaneous, and insects killed by this method preserve very well. It is especially useful for all kinds of insect larvæ.

The water should be quite hot, just hotter than you can bear, and the insect can be dipped into the water in between a folded piece of muslin.

PRESERVING. There are two useful liquids in which insects can be kept.

(*a*) Alcohol. 70°/₀ is the best strength for general use. It is rather expensive because "Alcohol 90°/₀" has to be diluted to the required percentage. On *no account* use methylated spirit as now sold in the shops : schools wishing to use the old sort of methylated spirit must get an Inland Revenue license.

(*b*) Formaldehyde (or " Formalin "). This is to be highly recommended both on the ground of cheapness and of efficiency. A weak solution (about 5 °/₀—7 °/₀) is used and a pint bottle of commercial formalin (costing 1*s*. 6*d*.) will make three or four gallons of preserving fluid. Inquire the price of formaldehyde.

Wide mouth bottles with good corks and 3″ by 1″ collecting tubes are most useful for storage of specimens. The tubes can be kept upright in cardboard or wooden boxes, and by labelling on the top of the cork as well as on the side of the tube any tube is easily found.

Lantern slides of insects.

Lantern slides are very useful, if used wisely, for summing up the results of the observational work or for explaining obscure points of structure or life-history. They must be kept strictly subordinate and not allowed to interfere with individual observational work.

There are many firms supplying excellent slides, e.g. :

Messrs Flatters and Garnett, 309, Oxford Road, Manchester.

Messrs Reynolds and Branson, Commercial Street, Leeds.

Messrs Newton and Co., 3, Fleet Street, London.

Messrs W. Watson and Son, 313, High Holborn, London.

The teacher who is keen upon Pond Life, or any natural history subject, should begin to get together a collection of useful slides made from his own or his friends' photographs. Photomicrography is comparatively easy and makes a very interesting winter hobby. A photomicrograph thrown on the screen will often serve the same purpose as showing a microscopic slide by means of a microscope, with the advantage of all seeing the object at once and thus saving a great deal of time.

The photomicrographs reproduced in this book were all taken with a cheap student's microscope and an ordinary bellows camera. There are several cheap and reliable books on Photomicrography[1].

Microscopic slides of insects.

These also form a necessary part of the material required for the foregoing lessons. They can be made very easily [see Appendix III] or bought from dealers. See list above.

[1] *Photomicrography for beginners*, 1*s*. *Handbook of Photomicrography*, Hind and Randles, Routledge, 7*s*. 6*d*.

APPENDIX II

APPARATUS

A good deal of the apparatus mentioned below could be dispensed with. I have used tumblers and white saucers instead of rectangular tanks and white bowls. Pie dishes make good aquaria.

Aquaria, etc.

1. The most useful are shallow, circular, white earthenware dishes, often sold in sets of three. Pie dishes will do.

2. Large rectangular glass tanks, made of thick glass cast all in one piece, can be got from electric supply firms. They are really accumulator tanks.

3. Most dealers stock large aquaria, but the prices are prohibitive, and the teacher might get one made much more cheaply.

4. Small rectangular tanks, necessary for observational work. The cheapest and yet quite efficient kind are rectangular moulded specimen jars, a convenient size would be 15 cms. by 10½ cms. by 3½ cms. and these can be got from Messrs Gallenkamp and Co., Ltd., Finsbury Square, London, for about 1s. each.

5. Saucers, odd white ones can be bought very cheaply. Paint saucers are very useful from Messrs Rowney or Winsor and Newton.

6. Petri dishes. See under Bacteriology in the price lists of apparatus dealers. Most useful size for observations upon living insects is 3″ diameter. By using cleaned off ¼-plate negatives with half Petri dishes, double the number of observation cases are made available.

Collecting apparatus, Nets, etc.

A great deal of this can be home made or made by a tinsmith

The net shown in Fig. 3 with which I have captured all the aquatic insects mentioned in this book, can be made for about 2s.

A small sixpenny coffee strainer fastened to a halfpenny bamboo is quite efficient.

Most dealers in Nature Study apparatus supply useful nets. Messrs Flatters and Garnett have lately put a cheap net on the market costing 1*s*. Illustrations of this net and also of one or two other types of useful nets have been kindly lent by the above firm.

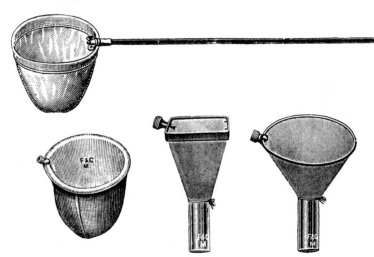

Fig. 47. Nets.

More elaborate outfits can be obtained; for details of these and of other useful collecting apparatus reference should be made to the catalogues of the undermentioned firms.

Messrs Flatters and Garnett, 309, Oxford Road, Manchester.

Messrs Watkin and Doncaster, Strand, London, W.C.

Messrs Reynolds and Branson, Commercial St., Leeds.

Messrs Philip Harris, 144, Edmund St., Birmingham.

Messrs A. Gallenkamp and Co., Ltd., Finsbury Sq., London, E.C.

From these firms can be obtained suitable apparatus for bringing home the spoils.

	s.	d.	
Glass tubes (3″ by 1″)	1	0	a dozen.
Wide-mouthed bottles (6—8 ozs.) ...	2	3	,,
Round tin glass-top boxes (3″ by 1½″)	2	6	,,

However these things are not essential, for all kinds of tobacco-tins are most efficient, and jam-jars on string are useful if the carrying distance is not great. See page 3 for hints upon the carriage of specimens.

Remember that most aquatic insects travel best in tin boxes with damp weed.

Microscopes, etc.

Pocket Lens. All instrument dealers stock a useful lens at 1*s.* each. Get the highest power possible in this type of lens.

The teacher should have a good aplanatic lens, both for his own use and also for use as a simple microscope to show details of structure to the class.

Microscope. Any reliable make of microscope will do. Beck's Star Microscope for £3. 15*s.* 0*d*. and Leitz School Microscope £5. 10*s.* 0*d*. can be highly recommended as good instruments at a reasonable figure.

Lantern Microscope. If a lantern is available it is often a great saving of time to show an object magnified by the microscope to the whole class at once. If only small magnification is desired it is possible to make a lantern microscope by means of a pocket lens in place of the front lens of the lantern. My first lantern microscope was fitted up with a pocket lens held by a piece of cork, with the microscopic slide of the object to be shown held by two pieces of bent wire, which were inserted in another flat piece of cork. By a few trials the best place for lens and slide can be found. Rather more rigid and precise apparatus is necessary for the use of higher lenses such as microscopic objectives, but with a little ingenuity and some skill in the use of tools a cheap and efficient lantern-microscope can be made.

It is important to have a good illuminant and the screen should be well shaded from any direct light.

Projection Microscope. Messrs Newton and Co., 3, Fleet Street, E.C., supply their patent microscopic attachment 8368 for £4. 14*s.* 6*d*. or 8369 for £3. 10*s.* 0*d*. Ordinary microscopic objectives can be used with this but Newton's specially constructed objectives give better definition up to the margin of the field on the screen.

Lantern. Most schools possess an optical lantern and, if used wisely, it is a most valuable assistant. A small screen is better than a large one—most lanternists use too large a screen—and for a class-room it will never be necessary to have one larger than 6 ft. square. The framework of the screen is most easily made with battens, and large sheets of white paper tacked on to this. A small screen such as this can generally be erected high up in a corner and then tilted slightly forwards so that all the class can see without difficulty. The lantern must also be tilted so that it points at right angles to the surface of the screen. It is quite a mistaken idea that the room need be completely darkened ; so long as the screen is well shaded from any direct light, it matters little about the rest of the room. Thus the class can work in comfort and take notes or make drawings with ease.

APPENDIX III

MICROSCOPICAL TECHNIQUE OR HOW TO MAKE MICROSCOPIC SLIDES

It has been thought advisable to give briefly the procedure for making temporary and permanent microscopic slides of such objects as occur in this book.

We have been dealing very largely with external features, and as mentioned in Chapter II the presence of chitin and its extraordinary resisting power make slide making of external features comparatively easy. If details of internal structure were to be studied, an elaborate procedure would be necessary, including fixing, staining, embedding in paraffin-wax or collodion and sectionizing in a microtome and then the careful and laborious examination of series of sections. Into this it is not necessary to go. It is, however, possible, by using small and transparent insects (such as young *Chironomus*, caddis and beetle larvæ) to see a good deal of the internal structure. The living insect is mounted whole in water and examined as a transparent object.

Temporary slides.

Things required. Glass slides or slips 3″ by 1″. A few excavated slips (these are slips with a round or oval depression for holding a thick object). Cover glasses, squares or circles (the circular shape is preferable and diameters of $\frac{3}{4}$″ and $\frac{7}{8}$″ are the most useful sizes). Mounting fluids such as water, dilute glycerine, glycerine jelly (these are best kept in bottles with a glass rod fixed to the stopper, such bottles can be bought from dealers). Mounted needles (these can be home-made with darning needles and elm twigs or penholders).

The object to be mounted is placed upon the centre of a glass slip with a drop of mounting fluid and a coverglass lowered on to this, so that the object is held between the two glasses, and the space round the object completely filled with the fluid to the exclusion of all air, for air bubbles often obscure the object.

Example. To make a temporary slide of the tail plates of a dragonfly larva to illustrate the principle of a tracheal gill:

1. Kill larva by plunging it into hot water.

2. Cut off the end of the abdomen with the three plates attached.

3. Mount in one of the mountants.

   (a) Water, very convenient but evaporates very quickly.

   (b) Glycerine, this does not evaporate like water and it makes the object slightly more transparent after a time.

Both of these will require care in manipulation so as not to displace the coverglass and so that no mounting fluid floods over on to the top of the coverglass. If this happens remove the coverglass, clean it and re-cover.

   (c) Glycerine jelly, at ordinary temperatures it is a jelly but it is easily made liquid and ready for use by standing the bottle in hot water. A bottle of glycerine jelly can be bought from dealers or can be made (see below).

Slides made with glycerine jelly are not permanent, but they can be made more so by painting a ring of some cement such as gold size or asphaltum round the edge of the coverglass.

Formula for a useful temporary mountant which makes the object more transparent than ordinary glycerine jelly :

Gilson's Chloral Hydrate Jelly.

1 volume of gelatin melted.

1 volume of Price's glycerine.

Mix together and add

1 volume of chloral hydrate. This is most easily done by adding crystals of chloral hydrate until the volume is increased by one half.

Warm until all the crystals are dissolved.

It is best kept in a Canada balsam bottle and should be placed in warm water some little time before it is needed for use.

Permanent slides.

Objects, which we wish to make into permanent slides, should be mounted in Canada balsam. The advantages of this mountant are its permanence, its perfect transparency, and its hardness when once set. The disadvantages are that it necessitates rather an elaborate procedure and the balsam takes a little time to set really hard, but the absolute permanence outweighs any disadvantages.

Things needed. Canada balsam dissolved in xylol, Canada balsam bottle ; xylol or benzole or oil of cloves or cedar oil ; absolute alcohol ; 90 °/₀ alcohol (methylated spirit must on no account be used); caustic soda or potash ; eosin, a red stain, dissolved in 90 °/₀ alcohol ; small porcelain evaporating dish ; spirit lamp or bunsen ; slides and covers ; mounted needles.

If we wish to make a permanent slide of the spiracle of a caterpillar showing tracheal tubes, or of some similar structure, the following procedure will be necessary, but it is frequently modified according to the nature of the object to be mounted.

1. Preparation of the object; e.g. take a caterpillar, freshly killed or preserved specimen, and cut off a piece of the body wall with at least two spiracles and tracheal tubes left on inside.

2. Heat gently in a 5 °/₀ solution of caustic potash. This is to get rid of muscle and other obscuring substances : it is made possible

by reason of the resisting power of chitin and the ease with which muscle can be dissolved out by an alkali.

Remove the object from the caustic from time to time into a saucer of water. The wooden end of a match is a convenient implement (metal corrodes in the hot alkali) and by a little pressure the dissolving muscle can be got away, and the progress of the operation watched.

3. When the specimen is transparent enough it should be washed in water made slightly acid and then in pure water.

Sometimes the above operations can be omitted when the specimen is sufficiently transparent and sometimes they have to be prolonged if a very dark and hard object is to be made at all transparent. One good way of making a very black object sufficiently transparent is to leave it in warm caustic all night. A small bottle containing caustic and specimen can be left on the kitchen grate or other warm place.

4. The specimen must now have all the water removed by means of increasing strengths of alcohol until it finally goes into absolute alcohol. This is a most important operation and one in which novices fail most, for if there is the slightest trace of water in the specimen anywhere (and of course if the specimen is freshly killed the tissues will contain a large amount of water) a whitish coloration will spread into the balsam and obscure the specimen.

> (i) If the specimen is fragile, thin and likely to become distorted if too quickly dehydrated, the following alcohols[1] should be used :
>
> 30 %, 50 %, 70 %, 90 %, absolute.
>
> (ii) If the specimen is not likely to suffer it can go at once into 90 % and then on to absolute.

If the object is very transparent it may be necessary to stain it so as to be able to see it properly. Eosin is the best stain for chitin and it is best used dissolved in 90 % alcohol. The specimen is then stained after its stay in the 90 % alcohol, and after the excess of stain has been washed out in a little more 90 %, it goes on to the absolute.

---

[1] These can easily be made by addition of water to 90 % alcohol.

The stain is a very rapid one, usually a minute or so is quite long enough.

5. The specimen has to be cleared by means of xylol or some similar substance. This removes the alcohol, permeates the specimen and prepares it for the balsam which is usually dissolved in xylol.

6. When the specimen looks clear and transparent it is placed on a clean slip and mounted in balsam.

The mount will need care for a few days until the xylol in the balsam has somewhat evaporated, and possibly a little more balsam may have to be run in under the coverglass to fill an air space caused by the consequent contraction of the balsam. This needs doing carefully so as not to get the balsam on to the top of the coverglass. It is best to drop a little balsam on to the slide, a little away from the coverglass, but on the side opposite to the air space, and then with a needle carefully move the balsam towards the coverglass; when it touches the edge of the cover the balsam should run under quite easily.

Summary of Method.

1. Preparation of object.

2. Remove soft parts with $5\,^\circ/_\circ$ caustic.

3. Wash in acid water and pure water.

4. Dehydrate in alcohols. Stain if necessary.

5. Clear in xylol or cedar oil.

6. Mount in balsam.

APPENDIX IV

BIBLIOGRAPHY

For classification and identification:

FURNEAUX. *Life in Ponds and Streams.* Longmans, Green. 6s.

R. LULHAM. *Introduction to Zoology (invertebrates).* Macmillan, 7s. 6d.

LYDDEKER. *Royal Natural History.* Vol. VI. *Invertebrates.* Warne.

L. C. MIALL. *The Natural History of Aquatic Insects.* Macmillan, 3s. 6d.

For reference and general reading:

LORD AVEBURY. *The Origin and Metamorphoses of Insects.* Macmillan, 3s. 6d.

G. H. CARPENTER. *The Life-story of Insects.* Camb. Univ. Press, 1s.

KIRBY AND SPENCE. *Introduction to Entomology.* Second-hand copies can be got.

O. H. LATTER. *Natural History of Common Animals.* Camb. Univ. Press, 5s.

L. C. MIALL. *Injurious and Useful Insects.* Bell, 3s. 6d.

——. "The Transformations of Insects." *Nature*, Vol. LIII.

L. C. MIALL AND HAMMOND. *The Harlequin Fly (Chironomus).* Clarendon Press, 7s. 6d.

D. SHARP. *The Cambridge Natural History*, Vols. V and VI. Macmillan, 17s. each.

The Book of Nature Study, Vol. II. Caxton Publ. Co., 7s. 6d.

Various papers in the *Transactions of Entomological, Linnean* and *Zoological Societies.* Reprints of such papers, and second-hand copies of many valuable books on insects can be got from Chamberlain and Co., Natural History Booksellers, Sheepscombe, Stroud, Gloucester, or from William Wesley and Son, 28 Essex Street, Strand.

INDEX

Printed in the United States
By Bookmasters